Wayne H

SAUNDERS GOLDEN SERIES

Mastering Chemistry

a problem solving guide for introductory chemistry

GERALD W. GIBSON

Department of Chemistry,
College of Charleston
Charleston, South Carolina

1975

W. B. SAUNDERS COMPANY • Philadelphia • London • Toronto

W. B. Saunders Company: West Washington Square
Philadelphia, PA 19105

12 Dyott Street
London, WC1A 1DB

833 Oxford Street
Toronto, Ontario M8Z 5T9, Canada

Mastering Chemistry   ISBN 0–7216–4108–3

© 1975 by W. B. Saunders Company. Copyright under the International Copyright Union. All rights reserved. This book is protected by copyright. No part of it may be reproduced, stored in a retrieval system, or transmitted in any form or by any means, electronic, mechanical, photocopying, recording, or otherwise, without written permission from the publisher. Made in the United States of America. Press of W. B. Saunders Company. Library of Congress catalog card number 74–12909

Last digit is the print number: 9   8   7   6   5   4   3   2   1

INTRODUCTION

What is the key to the mastery of chemistry?

All of us who teach chemistry are asked that question in one form or another at frequent intervals. There are two answers to it. The first: *motivation*. The second: *practice*.

Students come into an introductory chemistry course with a variety of goals and a disparity in depth of motivation. In a typical class, a few want to be research chemists, some are headed for medical school, some are budding biologists. All too seldom does anyone, regardless of his goals, really work at mastering chemistry. He may be highly motivated to "do well," but too often he has his real sights set on merely getting a good grade when they should be set on "getting good" at chemistry. There can be a difference. It is possible to achieve decent grades on quizzes without really working for mastery. It is much more likely, though, that one will be rated highly in something (that is, to "get good grades") if one is truly skilled at it. It makes sense, then, to study with this—becoming proficient at "doing chemistry"—as a goal rather than to study "for a quiz"; that is, rather than trying to figure out what is going to be asked on the test. Even if one ultimately succeeds in obtaining a good grade, his time is spent very inefficiently if in studying he has set the wrong goals for himself.

It is important, then, not only that one be highly motivated but also that one be properly motivated, and it is wise to examine at the outset what one is really trying to accomplish in chemistry.

The second key, practice, is what this book is designed to supply. It is strange that the same person who can so easily accept that to become good at, say, basketball, requires considerable practice will study chemistry by reading a textbook chapter and class notes. Reading is *not* practicing chemistry. Thoughtful reading can, of course, lead to comprehension of *concepts,* and one should strive for this by all means. A great portion of the discipline of chemistry, however, involves the *application* of concepts and principles—in other words, problem-solving. This is the part of chemistry that gives students the most trouble, not because they tend not to be bright enough to solve problems, but because they never really get any practice doing it except while taking a test.

The purpose of this problem-solving guide is to provide the motivated chemistry student with problems for practice, giving solutions to them and insight into approaches to solving them. If used properly, this method of studying chemistry gives a maximum pay-off for the time invested.

A central feature of this guide is the "OEIOU" (one equation in one unknown) approach to solving a problem. The student often is inclined to attack a problem "piecemeal"—that is, to solve for intermediate pieces of calculated data, then use them in other calculations until the final "answer" is obtained. This piecemeal technique tends to obscure the overall relationships among unknown and knowns in a problem and makes true mastery of problem-solving very difficult. In fact, it encourages one to memorize how the text solved a problem and to use the same method on a test. It is much better to get into the habit of *thinking* rather than memorizing. The OEIOU approach encourages one to think through each problem, developing it in a logical, orderly, mathematical way, postponing any use of numbers for as long as possible. The final OEIOU shows at a glance how the unknown may be found from the known because it contains recognizable symbols rather than numerical values peculiar only to that one specific problem. This serves to stress the importance of relationships in a way that merely multiplying and dividing numbers cannot do. Moreover, the OEIOU is in a form that makes it easy to do the final calculation by slide rule or calculator without having to stop and jot down intermediate values.

Not all the problems contained in the guide involve calculations; some involve the mastery of techniques and predictions based on concepts and generalizations: predicting formulas, balancing equations, predicting bond types and molecular geometries. Nor is every single type of problem that may be found in an introductory text included. The decisions as to what ought to be included were based on: (1) whether "mastery" of a given area required practice; and (2) whether the area or problem type was important enough at the introductory level to warrant its inclusion. An attempt has been made to keep the problems "realistic"; that is to keep them typical of situations that a real chemist might encounter in terms of data available and answers that might be sought.

Here are a few helpful hints about using the guide to best advantage:

(1) Use it regularly. Don't wait until the night before a test.

(2) Read each statement carefully, thinking as you read. Don't hesitate to go to a text for further amplification of concepts and definitions, if necessary.

(3) Keep the answers on the right covered with a sheet of paper while working the problems that are on the left. Write out your own solution to the problem, *then* check the solution given in the guide. Keep a record of which problems you can and cannot do. Go back later and try the ones you missed. Continue this until you are successful at (and understand) all of them.

(4) Do not write answers and notes on the question side of the page! Once you have done this the guide is of little use.

It is not uncommon that those who practice solving chemistry problems until they are good at it actually *enjoy* solving chemistry problems. If you don't believe that, try it and see!

ACKNOWLEDGMENT

The author wishes to express his gratitude to the College of Charleston Foundation which supported this work by providing a sabbatical semester during which the manuscript was completed. His thanks go, too, to the colleagues and students who made helpful suggestions during the progress of the work.

CONTENTS

1
USING NUMBERS 1

2
DERIVING EQUATIONS 31

3
THE ATOM 41

4
COMPOUNDS 66

5
GRAM ATOMIC WEIGHTS, MOLES AND FORMULAS .. 111

6
CHEMICAL EQUATIONS 132

7
HEAT FLOW DURING CHEMICAL REACTIONS 144

8
THE FIRST LAW OF THERMODYNAMICS 159

9
THE SECOND LAW OF THERMODYNAMICS 176

10
GASES 187

11
LIQUIDS 211

12
SOLIDS 219

13
SOLUTIONS 228

14
KINETICS 261

15
EQUILIBRIA INVOLVING GASES 277

16
PRECIPITATION EQUILIBRIA 304

17
ACIDS AND BASES 324

18
OXIDATION-REDUCTION PROCESSES 360

19
COMPLEX IONS 394

20
ORGANIC CHEMICALS 410

PART 1

USING NUMBERS

Statement 1

Numbers used in chemistry problems represent one of the following:

measured values—numbers obtained by making a measurement (or the result of calculations using values obtained by measurements); or

defined values—values that are known not because we measured something to find them, but because they are set **exactly** by definition.

Question 1

In which of these statements do the numbers represent measured values?

a. There are 416 trees in the orchard.

b. We used 2.5 cups of sugar for the cake.

c. A liter is equivalent to 1000 milliliters.

d. There are 50.23 milliliters of water in the flask.

> **Answer 1**
>
> measured values:
>
> a. 416 trees (obtained by counting)
>
> b. 2.5 cups (obtained by measuring volume)
>
> d. 50.23 ml (obtained by measuring volume)
>
> defined value:
>
> c. 1000 ml — The word "milliliter" is *defined* as 1/1000 of a liter.

Question 2

Select the measured values from the following statements:

a. There are 12 eggs per dozen.

b. The table is 1.86 meters long.

c. The nearest star is 4.3 light years away.

d. We ate 12 eggs for breakfast.

> **Answer 2**
>
> measured values:
>
> b. 1.86 meters
>
> c. 4.3 light years
>
> d. 12 eggs
>
> defined value:
>
> a. 12 eggs.
>
> Note that "12 eggs" was a measured value in (d), but not in (a). There are *always* exactly 12 eggs per dozen; no measurement is needed to tell us this.

Statement 2

Defined values are always **exact**.

Measured values are never exact (except for those obtained by counting) but have a **precision** determined by the type of instrument used to make the measurement. There is some uncertainty concerning the last digit of a measured value because the instrument used is never sensitive enough to measure beyond this lower limit of precision.

The precision of a measurement is revealed by the number of **significant figures** used to express it. To determine the number of significant figures in a value, count the total number of digits, including the final, uncertain digit. Do not count zeroes used only to fix the decimal point.

Defined (exact) values have an infinite number of significant figures.

Question 3

How many significant figures are used to express these values?

a. 42 meters

b. 1042 ml

> **Answer 3**
>
> a. two
>
> b. four

Question 4

How many significant figures are used to express these values?

a. 3.000 centimeters

b. 0.003 centimeter

Answer 4

a. four — the zeroes express precision.

b. one — the zeroes here are used only to fix the decimal point.

Question 5

How many significant figures are used to express these values?

a. 1.042 ml

b. 0.0421 g

c. 10.0042 g

Answer 5

a. four

b. three

c. six

Question 6

A crucible weighing 20.42 g on a triple-beam balance weighs 20.4195 g on an analytical balance. Which determination gives the greater precision?

Answer 6

The second, because it has 6 significant figures; the first has only 4 significant figures.

Question 7

A certain industrialist, I. M. Riche, reported his taxable income as $1.9 million. His actual income was $1,917,423.

Will the Internal Revenue Service be satisfied with his use of significant figures?

Answer 7

Not likely. His precision is much greater (7 significant figures) than his reported income (2 significant figures) indicates. The government would be losing the tax on $17,423.

Similarly, in chemistry one should always use the number of significant figures warranted by the precision of his measurement — no more and no less.

Question 8

How many significant figures are implied for the number of milliliters when we say: "A liter is equivalent to one thousand milliliters"?

Answer 8

An infinite number of significant figures. This is a *defined* value, so it is *exact*:

$$\text{one thousand ml} = 1000.00\ldots\text{ml}$$

Statement 3

Exponential notation is often used to avoid confusion over whether zeroes at the end of a value are significant. Exponential notation is also the only practical way of writing **very large numbers**. For instance:

$$10{,}000{,}000 = 10 \times 10 \times 10 \times 10 \times 10 \times 10 \times 10 = 10^7$$

meaning that 10 must be multiplied by itself 7 times to give the number 10,000,000; or 10 is said to be "raised to the seventh power."

Large numbers that cannot be expressed simply as ten raised to a power are translated into exponential notation by rewriting as in the following example, with the decimal placed by convention after the **first** digit:

Example

$$16{,}471 = 1.6471 \times 10{,}000 = 1.6471 \times 10^4$$

PART 1—USING NUMBERS   5

Question 9

Rewrite the value 10,000 fireflies using exponential notation.

> **Answer 9**
>
> 10^4 fireflies
>
> $10,000 = 10 \times 10 \times 10 \times 10 = 10^4$

Question 10

Rewrite the value 2,413 gumdrops in exponential notation.

> **Answer 10**
>
> 2.413×10^3 gumdrops
>
> $2,413 = 2.413 \times 1000 = 2.413 \times 10^3$
>
> By convention, the decimal is placed after the *first* digit; we don't express the value as 24.13×10^2 or as 241.3×10^1, although both of these expressions are technically correct.

Question 11

There were 222 million telephones in the world at the beginning of the year 1968. Express this value in exponential notation.

> **Answer 11**
>
> 2.22×10^8 telephones

Question 12

The oldest fossil flea found to date is 1.2×10^8 years old. Translate this exponential notation into the usual numerical representation.

> **Answer 12**
>
> 120,000,000 years

Statement 4

Very small numbers are also most conveniently expressed using exponential notation. For instance:

$$0.0000001 = \frac{1}{10,000,000} = \frac{1}{10 \times 10 \times 10 \times 10 \times 10 \times 10 \times 10} = 10^{-7}$$

Question 13

Rewrite the value 0.0001 centimeter exponentially.

Answer 13

10^{-4} cm

$$0.0001 = \frac{1}{10,000} = 10^{-4}$$

Question 14

Rewrite the value 0.002413 sec exponentially.

Answer 14

2.413×10^{-3} sec

$$0.002413 = 2.413 \times \frac{1}{1000}$$
$$= 2.413 \times 10^{-3}$$

Question 15

At 20° Celsius and one atmosphere of pressure the solubility of oxygen in water is 0.02208 g per liter. Express the solubility of oxygen in water by use of exponential notation.

Answer 15

2.208×10^{-2} g/L

PART 1—USING NUMBERS

Question 16

A lead atom weighs 3.44×10^{-22} g. Rewrite this number in nonexponential notation.

Answer 16

0.000 000 000 000 000 000 000 344 g (See how useful exponential notation can be!)

Statement 5

When a value is expressed exponentially, the number of significant figures will be simply the number of digits in the number multiplying the exponential term. The exponential term serves only to fix the decimal point.

When one encounters a value containing no decimal and the value is not expressed exponentially, it is impossible to ascertain whether any zeroes at the end are significant or not. This emphasizes the usefulness of exponential notation in conveying information about precision.

Question 17

How many significant figures are there in the following values?

a. 2.413×10^3 gumdrops

b. 2.22×10^8 telephones

Answer 17

a. four

b. three

Question 18

How many significant figures are there in the following values:

a. 2.208×10^{-2} g/L

b. 3.44×10^{-22} g

Answer 18

a. four

b. three

Question 19

How many significant figures are there in the following values:

a. 600 ml

b. 6.0×10^2 ml

c. 6.00 ml

Answer 19

a. We can't be sure. There are either one, two, or three.

b. Two. Note that by using exponential notation we can be *sure* of the precision, whereas in (a) we could not.

c. Three. The zeroes here are *not* used to fix the decimal point but must express precision.

Statement 6

When values are used in performing calculations, the final value ("answer") obtained will often contain a larger number of digits than those contained in one or more of the values used. The question of how much to "**round off**" arises.

When **multiplying** or **dividing**, as a general rule, round off the result to the **same number of significant figures** as are shown in the least precise value used in the calculation.

Note: When doing calculations that contain measured values, be sure to operate on the **units** of measurement (e.g., g, L, or sec) as well as the values, cancelling units when appropriate and keeping them in the numerator or denominator when appropriate. The kinds of unit found in a result can be a good clue as to whether the problem was set up properly.

Question 20

Express the result of the following calculation correctly:

$$3.00 \text{ g/ml} \times 1.207 \text{ ml} = ?$$

Answer 20

3.62 g

(*Not* 3.621 g. The final digit would be meaningless. Note that the "ml" terms cancel.)

PART 1—USING NUMBERS   9

Question 21

6.811 kilograms × 0.75 L/kg = ?

Answer 21

5.1 L

(*Not* 5.10825 L)

Question 22

3.00 g / 1.107 ml = ?

Answer 22

2.71 g/ml

Question 23

7.325 kilowatt hr/3.6 hr = ?

Answer 23

2.0 kilowatts

Statement 7

When **several** multiplication and/or division steps are necessary, it is convenient to round off all the values used to **one more significant figure** than is shown in the least precise value used **before** calculating. This saves time and gives an acceptable level of precision for the result.

Question 24

Round off the values in the following calculation appropriately, and express the final result correctly:

$$1.0815 \text{ calories/g} \times \frac{2.66 \text{ g/ml}}{12 \text{ ml}} = ?$$

Answer 24

$$1.08 \text{ cal/g} \times \frac{2.66 \text{ g/ml}}{12 \text{ ml}} = 2.4 \text{ cal}$$

There are only two significant figures in "12 ml," the least precise value, so we round off to 1.08 cal/g (three significant figures). The other value, 2.66 g/ml, already contains three significant figures.

Question 25

Round off the values in the following calculation appropriately, and express the final result correctly:

$$2.00000 \text{ L} \times \frac{273 \text{ degrees}}{305 \text{ degrees}} \times 1.8487 \text{ g/L} = ?$$

Answer 25

$$2.000 \text{ L} \times \frac{273 \text{ degrees}}{305 \text{ degrees}} \times 1.849 \text{ g/L} = 3.31 \text{ g}$$

Statement 8

When **adding** or **subtracting**, round off the result to the same number of digits to the right of the decimal place as are shown in the least precise value used in the calculation.

Question 26

Express the result of this calculation correctly:

$$28.64 \text{ ml} + 10.1 \text{ ml} + 8.33 \text{ ml} = ?$$

Answer 26

47.1 ml

(*Not* 47.07 ml)

PART 1—USING NUMBERS   11

Question 27

$$125.605 \text{ g} + 3.91 \text{ g} - 17.2 \text{ g} = ?$$

Answer 27

112.3 g

(*Not* 112.315 g)

Question 28

The total weight recorded on an unopened box of cornflakes was 15.2 g. After a bowl of cornflakes had been served, the box (with remaining contents) was found to weigh 12.682 g. What weight of cornflakes was taken from the box?

Answer 28

2.5 g

Statement 9

When multiplying or dividing by (or into) an exponential notation, if the nonexponential portion of the result has more than one digit before the decimal, or has a value of less than one, by convention it is **rewritten with one digit before the decimal**, increasing the exponential term by one power of ten for each place to the left or decreasing it by one power of ten for each place to the right that the decimal had to be shifted in rewriting.

Note that this rewriting does not change the overall value; it is merely a helpful convention.

Question 29

Express the results of the following calculations in exponential notation:

a. $2.0 \times 1.871 \times 10^8 = ?$
b. $6.0 \times 1.871 \times 10^8 = ?$

Answer 29

a. 3.7×10^8

(One digit is before the decimal, so no need to rewrite.)

b. 1.1×10^9

The multiplication gave a rounded-off result of 11×10^8. The decimal in the "11" term must be shifted to the *left* by one place in order for the result to have only one digit before the decimal; so 10^8 is increased to 10^9.

Question 30

Express the results of the following calculations in exponential notation:

a. 7.11×10^4 g/day $\times$ 6.0 days = ?

b. 6.9 kg / 8.5×10^3 rabbits = ?

Answer 30

a. 4.6×10^5 g

b. 8.1×10^{-4} kg/rabbit

Note that when we *decrease* 10^{-3} by one power of ten, it becomes 10^{-4}: $-3-1 = -4$.

Statement 10

When two or more exponential notations are **multiplied**, the exponents (powers) are added:

$$10^a \times 10^b = 10^{(a+b)}$$

Question 31

$10^8 \times 10^4 = ?$

Answer 31

10^{12}

$10^8 \times 10^4 = 10^{8+4} = 10^{12}$

Question 32

$10^2 \times 10^4 \times 10^3 = ?$

PART 1—USING NUMBERS   13

Answer 32

10^9

$10^2 \times 10^4 \times 10^3 = 10^{2+4+3} = 10^9$

Question 33

$10^2 \times 10^4 \times 10^{-3} = ?$

Answer 33

10^3

Question 34

$2.11 \times 10^2 \times 4.0 \times 10^3 = ?$

Answer 34

8.4×10^5

$2.11 \times 10^2 \times 2.0 \times 10^3 = (2.11 \times 4.0) \times (10^2 \times 10^3)$

$= 8.4 \times 10^5$

Question 35

$8.00 \times 10^8 \times 2.0 \times 10^4 = ?$

Answer 35

1.6×10^{13} (rather than 16×10^{12})

Statement 11

When exponential notations are **divided**, the exponent of the divisor term is **subtracted** from the exponent of the dividend term: $10^a/10^b = 10^{(a-b)}$

Question 36

$10^5 / 10^2 = ?$

Answer 36

10^3

$10^5 / 10^2 = 10^{5-2} = 10^3$

Question 37

$\dfrac{8.000 \times 10^9}{2.0 \times 10^4} = ?$

Answer 37

4.0×10^5

$\dfrac{8.000}{2.0} \times \dfrac{10^9}{10^4} = 4.0 \times 10^5$

Question 38

$\dfrac{7.1 \times 10^3 \times 6.9 \times 10^4}{1.5 \times 10^{12}} = ?$

Answer 38

3.3×10^{-4}

$\dfrac{7.1 \times 10^3 \times 6.9 \times 10^4}{1.5 \times 10^{12}} = \dfrac{4.9 \times 10^8}{1.5 \times 10^{12}} = 3.3 \times 10^{-4}$

Statement 12

To raise an exponential number to a **power** (e.g., to square it or cube it), **multiply** the exponent by the power: $(10^a)^b = 10^{(a \times b)}$

Question 39

$(10^3)^4 = ?$

PART 1—USING NUMBERS   15

Answer 39

10^{12}

$(10^3)^4 = 10^{3 \times 4} = 10^{12}$

Question 40

$(2.0 \times 10^2)^3 = ?$

Answer 40

8.0×10^6

$(2.0 \times 10^2)^3 = (2.0)^3 \times (10^2)^3 = 8.0 \times 10^6$

Statement 13

To take the **root** (e.g., square root or cube root) of an exponential notation, **divide** the exponent by the root:

$$\sqrt[b]{10^a} = 10^{a/b}$$

Question 41

$\sqrt[3]{10^{12}} = ?$

Answer 41

10^4

$\sqrt[3]{10^{12}} = 10^{12/3} = 10^4$

Question 42

$\sqrt{10^8} = ?$

Answer 42

10^4

16 PART 1—USING NUMBERS

Statement 14

To **add** or **subtract** two exponential notations, first **rewrite** so that both are expressed with the **same exponential factor**. The result will be the sum or difference of the non-exponential factor times the exponential factor. (Note: If necessary, rewrite the result to give only one digit before the decimal.)

Question 43

$(7.105 \times 10^{18}) + (4.33 \times 10^{16}) = ?$

Answer 43

7.148×10^{18}

$(7.105 \times 10^{18}) + (4.3 \times 10^{16}) = 7.105 \times 10^{18} + 0.0433$

$\times 10^{18} = 7.148 \times 10^{18}$

Question 44

$(1.65 \times 10^{-6}) - (6.2 \times 10^{-7}) = ?$

Answer 44

1.03×10^{-6}

$(1.65 \times 10^{-6}) - (6.2 \times 10^{-7}) = (1.65 \times 10^{-6}) - (0.62 \times 10^{-6})$

$= 1.03 \times 10^{-6}$

Question 45

$(7.595 \times 10^{9}) + (6.04 \times 10^{7}) + (2.1 \times 10^{8}) = ?$

Answer 45

7.86×10^{9}

Statement 15

The **common logarithm** of a number is simply the power to which 10 must be raised to give that number. Except for 1, 10, 100, 1000, and so forth, all numbers have logarithms that consist of a **characteristic** (an integer) and a **mantissa** (a decimal fraction).

To find the common logarithm of a number:

(a) Rewrite the number (if necessary) in exponential form, using the accepted convention.

(b) Find the logarithm of the nonexponential factor in a log table, on a slide rule, or with a calculator. This value is the **mantissa** of the final logarithm.

(c) The **characteristic** of the final logarithm is the exponent of ten in the exponential factor.

(d) Add the characteristic and the mantissa to obtain the logarithm of the original number.

This logarithm is customarily symbolized by "$\log_{10}$," which is read, "log to the base ten."

Note: Four-place log tables were used for the problems below.

Question 46

$\log_{10} 100 = ?$

Answer 46

2.000 ... (*exactly* 2, assuming all 3 digits in 100 are significant)

$\log_{10} 100 = \log_{10} (1 \times 10^2) = 0.00 \ldots + 2 = 2.00 \ldots$

Question 47

$\log_{10} 10{,}000 = ?$

Answer 47

4.000 ...

Question 48

$\log_{10} 71.6 = ?$

Answer 48

1.8549

$\log_{10} 71.6 = \log_{10} (7.16 \times 10^1) = 0.8549 + 1 = 1.8549$

Question 49

$\log_{10} 418 = ?$

Answer 49

2.6212

Question 50

$\log_{10} 0.000100 = ?$

Answer 50

$-4.000\ldots$

$\log_{10} 0.000100 = \log_{10} (1.00 \times 10^{-4}) = 0.00\ldots - 4$

$= -4.00\ldots$

Question 51

$\log_{10} 0.000716 = ?$

Answer 51

-3.1451

$\log_{10} 0.000716 = \log_{10} (7.16 \times 10^{-4}) = 0.8549 - 4$

$= -3.1451$

Note: Another method is to leave the log in the form $0.8549-4$ or $\bar{4}.8549$ rather than subtracting to give -3.1451. In chemistry it is usually advantageous to do the subtracting.

Question 52

$\log_{10} 0.00418 = ?$

Answer 52

-2.3788

Question 53

$\log_{10} 3.045 = ?$

Answer 53

0.4836

Note: If the log table lists only the log of 3.04 and 3.05, but *not* 3.045, we *estimate* the amount to add to log 3.04 to get log 3.045. In this case log 3.04 = 0.4829 and log 3.05 = 0.4843, a difference of 0.0014. Since 3.045 is $\frac{5}{10}$ of the way between 3.04 and 3.05, we take $\frac{5}{10} \times 0.0014 = 0.0007$. Then:

$0.4829 + 0.0007 = 0.4836$

Statement 16

The logarithm of the **product** of two numbers is equal to the **sum** of their individual logarithms:

$$\log_{10} (a \times b) = \log_{10} a + \log_{10} b$$

Question 54

$\log_{10} (6.47 \times 10^3) = ?$

Answer 54

3.8109

$\log_{10} (6.47 \times 10^3) = \log_{10} 6.47 + \log_{10} 10^3$

$= 0.8109 + 3 = 3.8109$

You can see that this is precisely what we did above when we rewrote numbers exponentially to obtain a mantissa and a characteristic.

Question 55

$\log_{10} (3.66 \times 17.4) = ?$

Answer 55

1.8040

Question 56

$\log_{10} (21.1 \times 115) = ?$

Answer 56

3.3850

Statement 17

The logarithm of the **quotient** of two numbers is equal to the logarithm of the dividend minus the logarithm of the divisor:

$$\log_{10} (a/b) = \log_{10} a - \log_{10} b$$

Question 57

$\log_{10} \dfrac{42.0}{5.00} = ?$

Answer 57

0.9242

$\log_{10} (42.0/5.00) = \log_{10} 42.0 - \log_{10} 5.00$

$= 1.6232 - 0.6990$

$= 0.9242$

Question 58

$\log_{10} \dfrac{825}{41.6} = ?$

Answer 58

1.2974

Question 59

$$\log_{10} \frac{32.5 \times 3.25}{14.7} = ?$$

Answer 59

0.8565

$$\log_{10} \frac{32.5 \times 3.25}{14.7} = \log_{10} 32.5 + \log_{10} 3.25 - \log_{10} 14.7$$

$$= 1.5119 + 0.5119 - 1.1673$$

$$= 0.8565$$

Statement 18

The logarithm of a number **raised to a power** is equal to the power times the logarithm of the number:

$$\log_{10}(a)^b = b \times \log_{10} a$$

Question 60

$$\log_{10}(5)^2 = ?$$

Answer 60

1.3980

$$\log_{10}(5)^2 = 2 \times \log_{10} 5 = 2 \times 0.6990 = 1.3980$$

Question 61

$$\log_{10}(710)^3 = ?$$

Answer 61

8.5539

Statement 19

The logarithm of the **root of a number** is equal to the logarithm of the number divided by the root:

$$\log_{10} \sqrt[b]{a} = \frac{\log_{10} a}{b}$$

Question 62

$\log_{10} \sqrt[3]{710} = ?$

Answer 62

0.9504

$$\log_{10} \sqrt[3]{710} = \frac{\log_{10} 710}{3}$$

$$= \frac{2.8513}{3}$$

$$= 0.9504$$

Question 63

$\log_{10} \sqrt{144} = ?$

Answer 63

1.0792

Statement 20

The **natural logarithm** of a number is the power to which a number called e, equal to 2.71828..., must be raised to give the number. The base e is widely used in the sciences because it appears in several natural laws.

The $\log_e$ of any number b can be found easily from this relationship:

$$\log_e b = 2.303 \log_{10} b$$

Question 64

$\log_e 100 = ?$

Answer 64

4.606

$\log_e 100 = 2.303 \log_{10} 100 = 2.303 \, (2.00 \ldots) = 4.606$

Question 65

$\log_e 710 = ?$

Answer 65

6.566

Statement 21

From the preceding definition of a logarithm, it follows that one might also define an **antilogarithm**—a number which has a given value as its logarithm. Knowing the logarithm of a number, we can find the number itself by a procedure that is the reverse of that used to find logarithms.

To find the **antilog of a positive log**, multiply the antilog of the mantissa (from log table, slide rule, or calculator) by the antilog of the characteristic.

Question 66

$\text{antilog}_{10} \, 3.4048 = ?$

Answer 66

2.54×10^3

$\text{antilog}_{10} \, 3.4048 = \text{antilog}_{10} \, 3 \times \text{antilog}_{10} \, 0.4048$

$= 10^3 \times 2.54 \text{ or } 2.54 \times 10^3$

Question 67

$\text{antilog}_{10} \, 8.5024 = ?$

PART 1—USING NUMBERS

Answer 67

3.18×10^8

Question 68

$\text{antilog}_e \ 1.500 = ?$

Answer 68

4.48

$\text{antilog}_e \ 1.500 = \text{antilog}_{10} \ (1.500/2.303)$

$= \text{antilog}_{10} \ 0.6513 = 4.48$

Statement 22

To find the **antilog of a negative log**, subtract the mantissa from 1, subtract 1 from the characteristic, and multiply the antilogs of the two values obtained.

Question 69

$\text{antilog}_{10} \ (-7.2857) = ?$

Answer 69

5.18×10^{-8}

$\text{antilog}_{10} \ (-7.2857) = \text{antilog}_{10} \ (0.7143-8)$

$= \text{antilog}_{10} \ (0.7143) \times \text{antilog}_{10} \ (-8)$

$= 5.18 \times 10^{-8}$

Question 70

$\text{antilog}_{10} \ (-3.4776) = ?$

Answer 70

3.33×10^{-4}

TESTING YOUR MASTERY—Part 1 Using Numbers

Question 71

Select from among the values mentioned in the following statements those which represent measured values:

a. She broke 6 test tubes before lunch.

b. There are 6 test tubes per half dozen.

c. He drinks 6.5 liters of water per week.

d. The basket is 25.6 cm in diameter.

> **Answer 71**
>
> Measured values:
>
> a. 6 test tubes
>
> b. 6.5 L
>
> d. 25.6 cm

Question 72

Express the value 6,286,577 ions exponentially.

> **Answer 72**
>
> 6.286577×10^6 ions

Question 73

Express the value 63 million flasks exponentially.

> **Answer 73**
>
> 6.3×10^7 flasks

Question 74

How many significant figures are expressed in these values?

a. 0.8815 g

b. 1.0741 kg

c. 2.7×10^5 erg

Answer 74

a. 4 significant figures

b. 5 significant figures

c. 2 significant figures

Question 75

Express these values exponentially.

a. 0.0000181 cm

b. 0.40001 ml

Answer 75

a. 1.81×10^{-5} cm

b. 4.0001×10^{-1} ml

Question 76

Select from among the following values the one which is measured with greatest precision:

a. 0.0847 g

b. 0.8477 g

c. 0.0848 g

Answer 76

0.8477 g

Question 77

$(1.74 \times 10^{-3}$ cm$) - (3.85 \times 10^{-5}$ cm$) = ?$

Answer 77

1.70×10^{-3} cm

Question 78

A small capillary tube weighs 2.7841×10^{-1} g. When filled with an unknown liquid it weighs 0.35954 g. How much does the liquid weigh?

Answer 78

8.113×10^{-2} g (or 0.08113 g)

Question 79

78.62 nights $\times$ 3426 stars/night $\times$ 3.77 planets/star = ?

Answer 79

1.02×10^6 planets

Question 80

What is the total mass of four samples of sodium chloride whose individual masses are found to be 1.0718 g, 0.475 g, 1.18 g, and 0.26119 g?

Answer 80

2.99 g

Question 81

$$\frac{6.37 \times 10^8 \text{ cars} \times 4.00 \text{ tires/car}}{2.184 \times 10^{-2} \text{ tire/dollar}} = ?$$

Answer 81

1.17×10^{11}

Question 82

$$\frac{7.681 \times 10^5 \text{ fairies}}{29 \text{ fairies/pinhead}} = ?$$

Answer 82

2.6 × 10^4 pinheads

Question 83

$(3.95 \times 10^6)^3$ = ?

Answer 83

6.16 × 10^{10}

Question 84

$\log_{10} (17.911)$ = ?

Answer 84

1.2531

Question 85

$(7.411 \times 10^5) \times (3.33 \times 10^3) \times (4.1 \times 10^{-8})$ = ?

Answer 85

1.01 × 10^2 (or 101)

Question 86

$\log_{10} (0.518)$ = ?

Answer 86

−0.2856

Question 87

$\log_e (0.518)$ = ?

Answer 87

−0.6577

Question 88

$\log_{10} \dfrac{217.6 \times 0.1889}{4.34} = ?$

Answer 88

0.9764

Question 89

$\log_{10} (18.6)^4 = ?$

Answer 89

5.0780

Question 90

$\log_{10} \sqrt[4]{18.6} = ?$

Answer 90

0.3173

Question 91

$\text{antilog}_{10} (6.4166) = ?$

Answer 91

2.61×10^6

Question 92

$\text{antilog}_{10} (-6.4166) = ?$

Answer 92

3.83×10^{-7}

Question 93

$\text{antilog}_e (6.4166) = ?$

Answer 93

6.11×10^2 (or 611)

PART 2
DERIVING EQUATIONS

Statement 1

The key to the mastery of problem-solving in chemistry is this: **Acquire the habit of attacking each problem in a logical, orderly fashion.** Avoid piecemeal problem-solving.

The orderly development of a problem typically involves working with mathematical equations. One usually begins with a **defining equation**, which states some fundamental relationship; then suitable substitutions of known quantities for unknowns are made; and the final result is

one equation in one unknown (OEIOU).

This OEIOU expresses the quantity sought (the unknown) in terms of quantities given (the knowns). The process of obtaining this OEIOU is called **derivation**. To be able to obtain a correct OEIOU is much more important to the student than is the mere ability to "get the right answer" (numerically) to a problem.

Once the OEIOU is obtained, using appropriate **symbols** for the knowns and the unknown, one may then substitute **numerical values** and do the calculation which gives the final "answer." **This step should always be postponed until the OEIOU has been derived.** Using symbols rather than numerical values often saves time and always makes it easier to see the relationships involved. More importantly, derivation gives a **general** equation—one that can be used for any similar problem regardless of the numerical values involved. There may be several equally good paths to the OEIOU, but if each step is done correctly, the final result will **always be the same**.

Summary of General Problem-Solving Approach

(1) List unknown and knowns in symbols (not in numbers).

(2) Write defining equation.

(3) Make substitutions if necessary (again using symbols, not numbers) until there is only one unknown in the equation.

(4) Rearrange to give OEIOU in its usual form—unknown only on the left, knowns on the right side of the equation.

(5) Substitute numerical values.

(6) Carry out calculations.

PART 2—DERIVING EQUATIONS

Question 1

The *density* of an object is equal to the ratio of its mass to its volume. Give the defining equation for density.

Answer 1

$$D = m/v$$

Question 2

A crucible weighing 9.662 g is partially filled with aluminum shot to give a new weight of 12.662 g. If the volume taken up by the shot is 1.107 ml, derive OEIOU from which the density of aluminum could be found.

Answer 2

$$D = \frac{m_2 - m_1}{v}$$

Defining equation:

$$D = m/v$$

(unknown) (unknown) (known)

Substitution:

$$D = \frac{g_2 - g_1}{m_1}$$

(unknown) (known)

Notice that we did not use the numerical values, because the problem does not require any calculation.

Question 3

The density of a certain fruit juice is 1.011 g/ml. Derive an equation that could be used to find the weight of a given volume of fruit juice.

Answer 3

$m = Dv$

Defining equation:

$$D = m/v$$

Rearranging:

$$m = Dv \text{ or } g = D \text{ ml}$$

The OEIOU here was obtained by simply rearranging the defining equation. In this case, the defining equation was actually one equation in one unknown, but it is good to get into the habit of having only the unknown on the left of the equation.

Question 4

A pycnometer weighs 5.000 g and has a volume of 5.000 ml. When filled with a certain liquid it weighs 15.000 g. Derive an equation for calculating the density of the liquid from these data.

Answer 4

$$D_{liq} = (g_2 - g_1)/ml_{liq}$$

Statement 2

A common type of problem encountered in chemistry involves converting a value measured using one system of measurement into a value expressed in units of another system. In the typical case, there is a **direct proportionality** between the two systems. For these, the conversion may be carried out using the relationship:

$$\#a = C.F. \times \#b$$

where "$\#a$" (read "the number of a") is the measurement in one system, "$\#b$" the value

PART 2—DERIVING EQUATIONS

in the other system, and "C.F." is a **conversion factor** relating them. The conversion factor contains the units a/b and may be read "a per b."

TABLE 2.1 SOME VALUES FOR CONVERSIONS

1 liter (L)	1.057 quarts
1 milliliter (ml)	0.0610 cubic inch
	0.001 liter
1 gram (g)	2.205×10^{-3} pound
	1000 milligrams
1 centimeter (cm)	0.3937 inch
1 calorie (cal)	4.184×10^7 ergs

Question 5

a. Derive an equation from which one can convert quarts to liters.

b. Convert 2.114 qt to liters.

Answer 5

a. $\#L = (L/\text{qt}) \times \#\text{qt}$

Defining equation:

$$\#a = \text{C.F.} \times \#b$$

Substitution:

$$\#L = \frac{L}{\text{qt}} \times \#\text{qt}$$

b. 2.000 L

OEIOU:

$$\#L = \frac{L}{\text{qt}} \times \#\text{qt}$$

$$= \frac{1.00\ldots L}{1.057\ \text{qt}} \times 2.114\ \text{qt}$$

$$= 2.000\ L$$

Question 6

a. Derive an equation from which one can convert pounds to grams.

b. Convert 2.00 lb to grams.

Answer 6

a. $\#g = (g/lb) \times \#lb$

b. 9.07×10^2 g

Question 7

a. Derive an equation from which one can convert inches to centimeters.

b. Convert 3.00 inches to centimeters.

Answer 7

a. $\#cm = (cm/in) \times \#in$

b. 7.62 cm

Question 8

A liquid has a density of 1.500 g/ml.

a. Derive an equation for calculating the density of the liquid in terms of lb/in^3.

b. Calculate the density of the liquid in lb/in^3.

Answer 8

a. $\#(lb/in^3) = \dfrac{(lb/g)}{(in^3/ml)} \times \#(g/ml)$

$\#a = \text{C.F.} \times \#b$

$\#\dfrac{lb}{in^3} = \dfrac{(lb/in^3)}{(g/ml)} \times \#(g/ml)$

$= \dfrac{lb}{g} \times \dfrac{ml}{in^3}$

Notice that in the final step we arranged the C.F. portion of the equation to give (lb/g)—in which *lb* and *g* are units that can be converted into each other—and (in^3/ml), in which in^3 and *ml* can be converted into each other. This step is not mathematically essential, but it helps in avoiding confusion.

b. 5.42×10^{-2} lb/in^3

$$\#(lb/in^3) = \frac{2.205 \times 10^{-3} \text{ lb}}{1.00 \ldots \text{ g}} \times \frac{1.00 \ldots \text{ ml}}{0.0610 \text{ } in^3} \times 1.500 \text{ g/ml}$$

Question 9

a. Derive an equation for converting the heat evolved during a chemical reaction from calories per gram of reactant to ergs per milligram of reactant.

b. Convert 4.00×10^3 cal/g into ergs/mg.

Answer 9

a. $\#(erg/mg) = \dfrac{erg/cal}{mg/g} \times (cal/g)$

b. 1.67×10^8 erg/mg

Statement 3

Some conversions from one unit of measurement into another involve relationships (or "functions") more complex than simple proportionalities—which is one of the dangers in automatically using the familiar "ratio-and-proportion" method of solving problems.

An example is the conversion from one of the **temperature units** into another. The most common temperature scales are the Celsius (or centigrade), the Fahrenheit, and the Kelvin (or absolute) scales. They are related by the following equations:

$$°F = 32 + 1.8°C$$

$$°K = 273 + °C$$

Note: The values 32, 1.8, and 273 are exact values.

Question 10

a. Derive an equation for converting from °F into °C.

b. Convert 98.6°F into °C.

Answer 10

a. $°C = \dfrac{°F - 32}{1.8}$

b. 37.0°C

Question 11

a. Derive an equation for converting °K into °C.

b. Convert 310°K into °C.

Answer 11

a. $°C = °K - 273$

b. 37°C

Question 12

a. Derive an equation for converting °F into °K.

b. Convert 32°F into °K.

Answer 12

a. $°K = \dfrac{°F + 459}{1.8}$

b. 273°K

TESTING YOUR MASTERY—Part 2 Deriving Equations

Question 13

The mass of 6.023×10^{23} gold atoms is 196.97 g. Derive an equation that could be used to find the mass (in grams) of a single atom of gold.

Answer 13

$$g_{Au} = \frac{\#\text{ g total}}{\#\text{ atoms}}$$

Question 14

A flask weighing 21.7864 g was filled with a quantity of gold powder, thereby increasing the total weight to 31.9411 g. If the volume of the flask was 25.48 ml, derive OEIOU which could be used to find the density of the gold powder.

Answer 14

$$\text{D powder} = \frac{(\text{g flask})_{filled} - (\text{g flask})_{empty}}{\text{vol. flask}}$$

Question 15

Liquid mercury releases 0.033 calorie per gram when it cools by 1.00°C (i.e., it has a "specific heat" of 0.033 cal/g °C).

Derive an equation that could be used to find the heat evolved (ΔH) when 20.0 g of mercury is cooled by 10.0°C.

Answer 15

$$\Delta H = \text{sp. ht.} \times \#g \times °C$$

Question 16

A metal has a density of 1.061 g/ml. Derive OEIOU for finding its density in lb/in^3.

Answer 16

$$\#(\text{lb/in}^3) = \frac{\text{lb}}{\text{in}^3} \times \frac{\text{ml}}{\text{g}} \times \#(\text{gm/ml})$$

Question 17

Derive an equation that can be used to convert ergs/mg to calories/g.

PART 2–DERIVING EQUATIONS

Answer 17

$$\#(cal/g) = \frac{cal/erg}{g/mg} \times \#(erg/mg)$$

Question 18

Derive an equation that can be used to convert the boiling point of lead in °F to its boiling point in °K.

Answer 18

$$°K = \frac{°F + 459}{1.8}$$

Question 19

According to Einstein's famous equation, $\Delta E = \Delta m\, c^2$, where ΔE represents the energy released by converting the mass, Δm, entirely into energy, and c is the velocity of light. Show the appropriate OEIOU for calculating the mass change necessary for producing 2.0×10^{20} erg of energy.

Answer 19

$$\Delta m = \Delta E / c^2$$

Question 20

Zinc shot weighing 1.071 g was added to a buret partially filled with water. Before adding the shot, the buret read 13.64 ml; after adding the shot, the buret read 13.79 ml. Derive OEIOU which could be used to calculate the density of zinc.

Answer 20

$$D = \frac{\#g}{ml_2 - ml_1}$$

Question 21

A beaker weighing 20.316 g is partially filled with "mossy" tin to give a new weight of 22.447 g. The volume of the tin was previously found to be 0.371 ml. What OEIOU could be used to calculate the density of tin?

Answer 21

$$D = \frac{g_2 - g_1}{ml}$$

Question 22

A liquid having a density of 1.128 g/ml and a 2.5 cm^3 piece of metal having a density of 1.361 g/ml are placed in a 50.00-ml flask. Show OEIOU which could be used to determine how many grams of the liquid can be placed in the flask.

Answer 22

$$\#g_{liq} = D_{liq}(ml_{flask} - ml_{metal})$$

PART 3
THE ATOM

Statement 1

An **atom** is the smallest unit of an element. Each pure element is composed of a unique type of atom, which is characteristic of that element, and contains no other type of atom.

Atoms are composed of a relatively massive nucleus containing a certain number, Z, of positively charged particles (**protons**) and some number of neutral particles (**neutrons**), with the same number of negatively charged particles (**electrons**) of very small mass outside the nucleus as there are protons in the nucleus.

Z is called the **atomic number** of the element composed of atoms of that type.

The **mass number**, A, of an element is the weight in "atomic weight units" (awu) (defined later in this section) rounded off to the nearest whole number.

A and Z can be obtained from any periodic chart of the elements. The mathematical relationship of Z and A is as follows:

$$Z + \#\text{neutrons} = A$$

Question 1

A proton weighs 1.00728 atomic weight units, a neutron 1.00867 awu, and an electron 0.00055 awu. Give the mass numbers of these particles.

> #### Answer 1
>
> proton: 1
> neutron: 1
> electron: 0
>
> Round each to nearest whole number.

Question 2

A certain atom contains 16 protons and 18 neutrons. What is its mass number?

Answer 2

34

$$A = Z + \#\text{neutrons}$$

OEIOU: $A = \#\text{protons} + \#\text{neutrons}$

$$= 16 + 18 = 34$$

Question 3

A certain atom has a mass of 13.003 awu. Its atomic number is 6. How many neutrons does its nucleus contain?

Answer 3

7

$$A = Z + \#\text{neutrons}$$

OEIOU: $\#\text{neutrons} = A - Z$
$$= 13 - 6$$
$$= 7$$

Statement 2

In nature one finds atoms that have the **same atomic number**, but **different mass numbers**. These are said to be **isotopes** of the same element. An isotope may be specified by writing the symbol for the element, a hyphen, and then the mass number; or by writing the symbol for the element with the mass number as a superscript to the left.

Question 4

The two principal isotopes of carbon that occur in nature have masses of 12.000 awu and 13.0033 awu. Designate these two isotopes by appropriate symbols.

Answer 4

^{12}C and ^{13}C or C - 12 and C - 13

PART 3—THE ATOM

Question 5

Designate symbolically the isotopes of hydrogen that have the masses 1.0078 awu, 2.0141 awu, and 3.0160 awu.

Answer 5

^{1}H and ^{2}H and ^{3}H

or

H - 1 and H - 2 and H - 3

Statement 3

The **atomic weight** of an element is the mass of an **average atom** of the element occurring in nature. The mathematical formula is:

$$\text{atomic weight} = f_1 A_1 + f_2 A_2 + \ldots$$

where $f_1, f_2, \ldots$ are the fractional abundances of isotopes having the masses $A_1, A_2, \ldots$.

This convention is used because in everyday life we never deal with one atom at a time, but with vast numbers of them, and in doing ordinary calculations it is the average mass that is important.

Question 6

In a natural sample of carbon, 0.989 of it is composed of ^{12}C (12.0000 awu) and 0.0110 of it of ^{13}C (13.0033 awu). What is the atomic weight of carbon?

Answer 6

12.0

$$\text{atomic weight} = f_{12} A_{12} + f_{13} A_{13}$$

$$= (0.989)(12.000)$$

$$+ (0.110)(13.0033)$$

$$= 12.011$$

rounded off to the number of significant figures warranted by the data as given: 12.0.

Question 7

Natural oxygen consists of 99.76 per cent ^{16}O (15.9949 awu), 0.04 per cent ^{17}O (16.9991 awu), and 0.20 per cent ^{18}O (17.9992 awu). Assuming the percentages to be exact values, what is the atomic weight of oxygen?

Answer 7

15.9993

$$\text{atomic weight} = f_{16}A_{16} + f_{17}A_{17} + f_{18}A_{18}$$

$$= (0.9976)(15.9949)$$

$$+ (0.0004)(16.9991)$$

$$+ (0.0020)(17.9992)$$

$$= 15.9993$$

Question 8

Nitrogen has an atomic weight of 14.0067 awu. Its naturally occurring isotopes are ^{14}N (14.0031) and ^{15}N (15.0001 awu). What are the relative abundances of each isotope?

Answer 8

0.9964 is ^{14}N.
0.0036 is ^{15}N.

Defining equation:

$$\text{at. wt.} = f_{14}A_{14} + f_{15}A_{15}$$

We also know:

$$f_{14} + f_{15} = 1.000 \ldots$$

(The sum of all the fractions must be unity.)

Therefore:

$$f_{14} = 1.0000 - f_{15}$$

Substituting:

$$\text{at. wt.} = (1.0000 - f_{15})A_{14} + f_{15}A_{15}$$

OEIOU:

$$f_{15} = \frac{\text{at. wt.} - A_{14}}{A_{15} - A_{14}}$$

$$= \frac{14.0067 - 14.0031}{15.0001 - 14.0031}$$

$$= 0.0036$$

Then (OEIOU):

$$f_{14} = 1.0000 - f_{15}$$

$$= 0.9964$$

Statement 4

The **atomic weight unit** is defined as exactly $\frac{1}{12}$ the mass of an atom of ^{12}C. This means that the ^{12}C isotope is the standard against which all other atomic weights are determined; it is assigned a mass of exactly 12.000. . . .

Question 9

An average sulfur atom is 2.672 times as massive as an atom of ^{12}C. What is the atomic weight of sulfur?

Answer 9

32.06 awu

$$\text{mass } ^{12}C \times \text{factor} = \text{at. wt. S}$$

$$\text{at. wt. S} = 12.000 \times 2.672$$

$$= 32.06$$

Question 10

The principal isotope of sulfur is 2.664 times as massive as an atom of ^{12}C.

a. What is its mass?

b. Which isotope of sulfur is it?

> ### Answer 10
>
> a. 31.97 awu
>
> b. ^{32}S

Statement 5

The modern model of the atom, known as the "**wave-mechanical**" **model**, proposes that an electron in an atom is in some ways "wavelike" (continuous) and in other ways "particle-like" (discrete). It is not the same sort of hard object as a billiard ball, for example, and doesn't act as a billiard ball appears to us to act. It is a vague, tenuous uncertain sort of thing for a piece of matter to be—but experiments performed to the present time support this description.

We can not know, according to the Heisenberg Uncertainty Principle, both the position and the velocity of an electron simultaneously. We cannot, therefore, speak of an electron "orbiting" a nucleus. However, using Schrödinger's wave equations (please refer to a good text for details about this equation and for "pictures" of the orbitals described below) we can calculate (for simple systems) the probability of finding an electron in a given region near the nucleus. The region in which the probability is greatest is called an **orbital**. The orbital in which an electron is most likely to be found is defined by the **quantum numbers** for the electron. The orbital may be spherical (s orbital), "dumbbell" shaped (p orbital), or have other more complicated shapes (d orbital, f orbital, and so forth).

The electrons in an atom may vary in energy depending on their orbital conditions. Only certain **specific** energy values are possible, with each of these allowable energies being fixed by a set of **quantum numbers**.

TABLE 3-1 QUANTUM NUMBERS

Symbol	Name	Designates	Allowable Values
n	principal	size of orbital (level or shell)	1, 2, 3, ...
l	angular momentum	shape of orbital (sublevel or subshell)	0, 1, 2, ... (n–1)
m_l	magnetic	orientation of orbital (orbital)	$l, l-1, l-2, \ldots 0, -1, -2, \ldots$
m_s	spin	spin direction of electron	$+\frac{1}{2}, -\frac{1}{2}$ (for each value of m_l)

The total number of electrons that may have a given principal quantum number (or be in a given "level") is equal to $2n^2$.

Question 11

What angular momentum quantum numbers are possible in a level for which $n = 1$?

Answer 11

$l = 0$ (zero)

Because $n-1$ is the highest allowable value and $n-1 = 0$ when $n = 1$, the only allowable value for l here is zero.

Question 12

What angular momentum quantum numbers are possible in the $n = 2$ level?

Answer 12

$l = 0$ and 1

The highest allowable l value is $n - 1$, which in this case is $2 - 1 = 1$. Any positive integer below this value is also allowed, so the other allowed value is 0.

Question 13

What angular momentum quantum numbers are possible in the $n = 3$ level?

Answer 13

$l = 0, 1,$ and 2

Question 14

What magnetic quantum numbers are possible when $l = 0$?

Answer 14

$m_l = 0$

Because the maximum value allowed is l and the minimum value allowed is $-l$, the only allowable value for m_l is zero.

Question 15

What m_l values are possible when $l = 1$?

Answer 15

$m_l = 1, 0,$ and -1

Values are allowed to range from l to $-l$. Since $l = 1$, $l-1=0$, and $l-2=-1$; these are the permitted values for m_l.

Question 16

What m_l values are permitted when $l = 2$?

Answer 16

$m_l = 2, 1, 0, -1,$ and -2

Question 17

What m_s values are permitted when $m_l = 0$?

Answer 17

$+\frac{1}{2}$ and $-\frac{1}{2}$

Question 18

What m_s values are allowed when $m_l = 1$?

Answer 18

$+\frac{1}{2}$ and $-\frac{1}{2}$

These are the *only* possible values, regardless of what the other quantum numbers are.

Question 19

What total number of electrons may there be in a level having $n=1$?

Answer 19

2

Total number of electrons = $2n^2$

$= 2(1)^2 = 2$

Question 20

What total number of electrons can there be in a level having

a. $n = 2$?

b. $n = 3$?

c. $n = 4$?

Answer 20

a. 8

b. 18

c. 32

Question 21

a. How many occupied orbitals in the Ar atom have $m_l = -1$ as their quantum number?

b. How many electrons in the Ar atom have $m_l = -1$ as one of their quantum numbers?

Answer 21

a. 2 orbitals

b. 4 electrons

For Ar, Z=18, so Ar has a total of 18 electrons.

TABLE 3-2   SUMMARY OF ALLOWED ORBITALS AND QUANTUM NUMBERS

n	l	m_l	m_s
1	0	0	$+\frac{1}{2}, -\frac{1}{2}$
2	0	0	$+\frac{1}{2}, -\frac{1}{2}$
	1	1	$+\frac{1}{2}, -\frac{1}{2}$
		0	$+\frac{1}{2}, -\frac{1}{2}$
		−1*	$+\frac{1}{2}**, -\frac{1}{2}**$
3	0	0	$+\frac{1}{2}, -\frac{1}{2}$
	1	1	$+\frac{1}{2}, -\frac{1}{2}$
		0	$+\frac{1}{2}, -\frac{1}{2}$
		−1*	$+\frac{1}{2}**, -\frac{1}{2}**$
	2†	2†	$+\frac{1}{2}†, -\frac{1}{2}†$
		1†	$+\frac{1}{2}†, -\frac{1}{2}†$
		0†	$+\frac{1}{2}†, -\frac{1}{2}†$
		−1†	$+\frac{1}{2}†, -\frac{1}{2}†$
		−2†	$+\frac{1}{2}†, -\frac{1}{2}†$

*These two orbitals are occupied and have $m_l = -1$.
**These four electrons have $m_l = -1$ as one of their quantum numbers.
†These orbitals are unoccupied.

Statement 6

When stating the orbital conditions of an electron it is convenient to use a shorthand notation that specifies the principal and angular momentum quantum numbers (hence the size and shape of the orbital).

For ease in reading, the actual quantum numbers are replaced by letters taken from **spectroscopic** terms for the angular momentum quantum number:

$$s \text{ is used for } 0$$
$$p \text{ for } 1$$
$$d \text{ for } 2$$
$$f \text{ for } 3$$

This combination of principal quantum number with a letter to designate the angular momentum quantum number is called the **spectroscopic notation** of the orbital or electron.

Example

$$n = 2 \text{ and } l = 0$$
$$\text{spectroscopic notation: } 2s$$

Question 22

Give the spectroscopic notation for an electron with $n = 3$ and $l = 1$.

Answer 22

$3p$

Question 23

Give the spectroscopic notation for an electron with $n = 4$ and $l = 2$.

Answer 23

$4d$

Statement 7

The orbital locations of all electrons in an atom may also be described using spectroscopic notation. This description is called the **electron configuration** of the atom. It consists of the spectroscopic notation for each occupied sublevel plus a superscript indicating the number of electrons in the sublevel.

The steps for deciding on the correct electron configuration for an element are as follows:

a. Decide on the **total** number of electrons to be placed in orbitals;

b. Write out the spectroscopic notation for each sublevel in order of increasing energy: 1s 2s 2p 3s 3p 4s 3d 4p 5s 4d 5p 6s 4f 5d 6p 7s . . .

c. Using **superscripts**, allot electrons to each sublevel in appropriate numbers until **all** of the electrons have been allocated. The appropriate number of electrons that may be accommodated (as a maximum) in the sublevels are as follows:

$$2 \text{ in the } s \text{ sublevel}$$
$$6 \text{ in the } p \text{ sublevel}$$
$$10 \text{ in the } d \text{ sublevel}$$
$$14 \text{ in the } f \text{ sublevel}$$

This scheme for arriving at a probable electron configuration for an atom is often called an **aufbau** (building up) **process**.

Question 24

What is the electron configuration for carbon?

Answer 24

C (6 electrons) $1s^2 \ 2s^2 \ 2p^2$

Question 25

What is the electron configuration for silicon?

Answer 25

Si (14) $1s^2 \ 2s^2 \ 2p^6 \ 3s^2 \ 3p^2$

Question 26

What is the electron configuration for titanium?

Answer 26

Ti (22) $1s^2 \ 2s^2 \ 2p^6 \ 3s^2 \ 3p^6 \ 4s^2 \ 3d^2$

Question 27

What is the electron configuration for hafnium?

Answer 27

Hf (72) $1s^2\ 2s^2\ 2p^6\ 3s^2\ 3p^6\ 4s^2\ 3d^{10}\ 4p^6\ 5s^2\ 4d^{10}\ 5p^6\ 6s^2$
$4f^{14}\ 5d^2$

Question 28

What family of elements on the periodic table has the electron configuration $ns^2\ np^5$, where n is the highest principal quantum number in each element?

Answer 28

The halogens in Group 7A.

$n = 2$ for F
$n = 3$ for Cl
$n = 4$ for Br
$n = 5$ for I

Statement 8

Orbital diagrams give an even more detailed revelation of electron configuration, including the **specific orbitals** within a sublevel that are occupied and the spins of the electrons in them. Boxes are used to represent individual orbitals, and arrows indicate when $m_s = +\frac{1}{2}(\uparrow)$ or $-\frac{1}{2}(\downarrow)$.

In assigning electrons to orbitals, we are guided by these two rules:

Pauli Exclusion Principle Not more than two electrons may be placed in one orbital.

Hund's Rule When orbitals of equal energies (in the same sublevel) are available, one electron is placed in each orbital before placing 2 electrons in any one orbital.

Example: N(7)

	$1s^2$	$2s^2$	$2p^3$
spectroscopic notation			
orbital diagram	[↑↓]	[↑↓]	[↑ ↑ ↑]
m_l values	0	0	+1 0 -1

Question 29

Draw an orbital diagram for fluorine.

Answer 29

F(9) $1s^2$ $2s^2$ $2p^5$

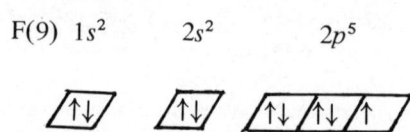

Question 30

Draw an orbital diagram for manganese.

Answer 30

Mn(25) $1s^2$ $2s^2$ $2p^6$ $3s^2$

$3p^6$ $4s^2$ $3d^5$

Question 31

Draw an orbital diagram for the two highest-energy orbitals in hafnium.

Answer 31

$4f^{14}$ $5d^2$

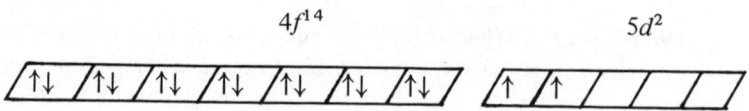

Statement 9

According to the Pauli exclusion principle, an orbital may contain either one or two electrons. If there are two electrons in an orbital, they will be **paired**; that is, they will have opposite spins ($+\frac{1}{2}$ and $-\frac{1}{2}$). If there is only one electron in an orbital, it is considered **unpaired**.

Elements having one or more unpaired electrons are **paramagnetic** (attracted into a magnetic field); those having only paired electrons are **diamagnetic** (repelled by a magnetic field).

Orbital diagrams can be very useful in predicting whether an element will be paramagnetic or diamagnetic.

Question 32

How many unpaired electrons are there in an oxygen atom?

Answer 32

2

orbital diagram:

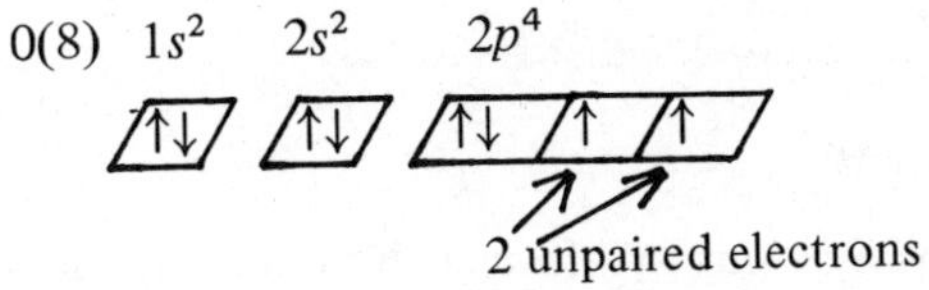

O(8) $1s^2$ $2s^2$ $2p^4$

[↑↓] [↑↓] [↑↓][↑][↑]

2 unpaired electrons

Question 33

Is the oxygen atom diamagnetic or paramagnetic?

Answer 33

paramagnetic

From the orbital diagram (see Answer 32) we can see that there are two unpaired electrons; thus the atom is paramagnetic.

Question 34

Would you expect these atoms to be diamagnetic or paramagnetic?

a. Ne

b. Fe

Answer 34

a. diamagnetic

Ne(10) $1s^2$ $2s^2$ $2p^6$

[↑↓] [↑↓] [↑↓] [↑↓] [↑↓]

All electrons are paired.

b. paramagnetic

Fe(26) $1s^2$ $2s^2$ $2p^6$ $3s^2$

[↑↓] [↑↓] [↑↓] [↑↓] [↑↓] [↑↓]

$3p^6$ $4s^2$ $3d^6$

[↑↓] [↑↓] [↑↓] [↑↓] [↑↓] [↑] [↑] [↑] [↑]

Four unpaired electrons

Statement 10

When **positive ions** are formed by removing one or more electrons from an atom, electrons are removed in the following order:

$$np$$
$$ns$$
$$(n-1)\,d$$
$$(n-2)\,f$$

where n is the principal quantum number of the **highest** occupied level; $(n-1)$ is the second highest; and $(n-2)$ is the third highest.

When **negative ions** are formed, electrons are placed in the next available orbital, following the normal aufban procedure.

Question 35

Show the electron configuration for the Na^+ ion.

Answer 35

$Na^+(10)$ $1s^2$ $2s^2$ $2s^6$

The Na atom has one electron in the 3s orbital, its highest occupied orbital. This electron is removed, leaving the configuration shown.

PART 3—THE ATOM

Question 36

Show the electron configuration for the F^- ion.

Answer 36

F^- (10) $1s^2$ $2s^2$ $2p^6$

Question 37

Show the electron configuration for the Zn^{2+} ion.

Answer 37

Zn^{2+} (28) $1s^2$ $2s^2$ $2p^6$ $3s^2$ $3p^6$ $3d^{10}$

Notice that although we filled the $3d$ sublevel *last*, the $4s$ electrons are removed *first*.

Question 38

Show the electron configuration of the Ca^{2+} ion.

Answer 38

Ca^{2+} (18) $1s^2$ $2s^2$ $2p^6$ $3s^2$ $3p^6$

Question 39

Show the electron configuration and orbital diagram for the Fe^{3+} ion.

Answer 39

Fe^{3+} (23) $1s^2$ $2s^2$ $2p^6$ $3s^2$

[↑↓] [↑↓] [↑↓][↑↓][↑↓] [↑↓]

$3p^6$ $4s^0$ $3d^5$

[↑↓][↑↓][↑↓] [ ] [↑][↑][↑][↑][↑]

Question 40

Show the electron configuration and orbital diagram for the O^{2-} ion.

Answer 40

O^{2-} (10) $1s^2$ $2s^2$ $2p^6$

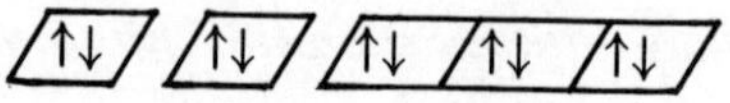

| ↑↓ | ↑↓ | ↑↓ | ↑↓ | ↑↓ |

Statement 11

An atom or ion may absorb radiation and become **excited** (i.e., an electron is promoted to a higher energy level). An unexcited atom is said to be in its **ground state**. In the **first excited state** of an atom or ion, one electron has been promoted from the highest occupied atomic orbital to the lowest unoccupied atomic orbital.

Question 41

Show the electron configuration for the first excited state of Cl.

Answer 41

Cl(17) first excited state:

$1s^2$ $2s^2$ $2p^6$ $3s^2$ $3p^4$ $4s^1$

Question 42

Show the electron configuration for the first excited state of the F^- ion.

Answer 42

F^- (10) first excited state:

$1s^2$ $2s^2$ $2p^5$ $3s^1$

TESTING YOUR MASTERY—Part 3 The Atom

Question 43

What is the mass number of an atom having an atomic number of 41 and 52 neutrons?

Answer 43

93

OEIOU: $A = Z + \#\,neutrons$

Question 44

An atom with an atomic number of 17 has a mass of 34.96885 awu. How many neutrons does its nucleus contain?

Answer 44

18

OEIOU: $\#\,neutrons = A - Z$

Question 45

An atom has a mass of 14.0031 awu and contains 7 neutrons. What is the atomic number of this element?

Answer 45

7

OEIOU: $Z = A - \#\,neutrons$

Question 46

Designate symbolically the isotopes of the element having an atomic number of 16 and masses of 31.972 awu, 32.971 awu, 33.968 awu, and 35.967 awu.

Answer 46

^{32}S, ^{33}S, ^{34}S, and ^{36}S

or

S-32, S-33, S-34, and S-36

The atomic number tells us which element this is.

Question 47

Naturally occurring potassium consists of 93.10 per cent ^{39}K (38.98 awu), 6.88 per cent ^{41}K (40.97 awu) and 0.01 per cent ^{40}K (39.98 awu). Find the atomic weight of potassium to the proper number of significant figures.

Answer 47

39.1 awu

OEIOU: at. wt. $= f_{39} A_{39} + f_{40} A_{40} + f_{41} A_{41}$

Question 48

If hydrogen has an atomic weight of 1.0080, and if its naturally occurring isotopes are ^{1}H (1.0078) and ^{2}H (2.0141), what are the relative abundances of each isotope?

Answer 48

$f_1 = 0.9998$

$f_2 = 0.0002$

OEIOU: $f_1 = \dfrac{\text{at. wt.} - A_2}{A_1 - A_2}$

OEIOU: $f_2 = 1.000 \ldots - f_1$

Question 49

The principal isotope of chlorine is 2.9140 times as massive as an atom of ^{12}C.

a. What is its atomic mass?

b. What isotope of chlorine is it?

PART 3—THE ATOM

Answer 49

a. 34.969 awu

b. ^{35}Cl

Question 50

a. What angular momentum quantum numbers are allowed when $n = 4$?

b. How many subshells are allowed when $n = 4$?

Answer 50

a. $l = 0, 1, 2,$ and 3

b. 4 subshells, each corresponding to one of the four l values in (a).

Question 51

Give the identifying spectroscopic notation for each of the four subshells (or sublevels) that may be occupied when $n = 4$.

Answer 51

4s 4p 4d 4f

Question 52

What m_l values are allowed for each of the subshells that may be occupied when $n = 4$?

Answer 52

For 4s, $m_l = 0$

For 4p, $m_l = 1, 0, -1$

For 4d, $m_l = 2, 1, 0, -1, -2$

For 4f, $m_l = 3, 2, 1, 0, -1, -2, -3$

PART 3—THE ATOM

Question 53

a. What total number of electrons may an atom have in a shell (level) for which $n = 4$?

b. What total number of electrons may an atom have if its maximum value of $n = 4$?

Answer 53

a. 32

OEIOU: $\#e = 2n^2$

b. 60

OEIOU: total number of elements = $(2n^2)_1 + (2n^2)_2 + (2n^2)_3 + (2n^2)_4$

Question 54

What are the allowed m_s values when $m_l = 3$?

Answer 54

$+\frac{1}{2}$ and $-\frac{1}{2}$

Question 55

a. How many occupied orbitals in the Al atom have $m_l = 0$ as a quantum number?

b. How many electrons in the Al atom are in s orbitals?

Answer 55

a. 4 orbitals

b. 6 electrons in s orbitals

Question 56

Give the spectroscopic notation for an electron having this set of quantum numbers: $n = 3, l = 2, m_l$ and $m_s = +\frac{1}{2}$.

PART 3—THE ATOM

Answer 56

3d

Note: Additional letter subscripts may be used to distinguish among the several possible orientations (m_l values) for p, d, f, etc. orbitals, but these are often omitted at the introductory level.

Question 57

What is the electron configuration for the magnesium atom?

Answer 57

Mg(12) $1s^2\ 2s^2\ 2p^6\ 3s^2$

Question 58

Show the predicted electron configuration for the indium atom.

Answer 58

In (49) $1s^2\ 2s^2\ 2p^6\ 3s^2\ 3p^6\ 4s^2\ 3d^{10}\ 4p^6\ 5s^2\ 4d^{10}\ 5p^1$

Question 59

What family of elements on the periodic table has the electron configuration $(n-1)s^2\ (n-1)p^6\ ns^1$, where n is the highest principal quantum number in each element?

Answer 59

The alkali metals in Group 1A.

Question 60

Draw the orbital diagram for sodium.

Answer 60

Na(11) $1s^2$ $2s^2$ $2p^6$ $3s^1$

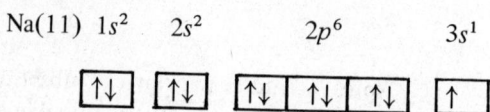

Question 61

Draw the orbital diagram for the two highest-energy orbitals in vanadium.

Answer 61

$4s^2$ $3d^3$

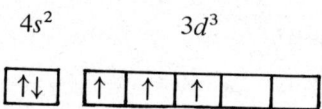

Question 62

Show the orbital diagram for the two highest-energy orbitals of tungsten.

Answer 62

$4f^{14}$ $5d^4$

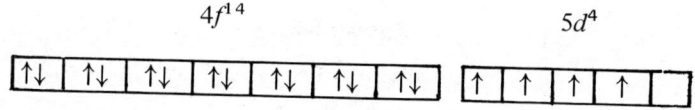

Question 63

Show the orbital diagram for the S^{2-} ion.

Answer 63

$S^{2-}(18)$

$1s^2$ $2s^2$ $2p^6$ $3s^2$ $3p^6$

Question 64

Draw the orbital diagram for the V^{3+} ion.

Answer 64

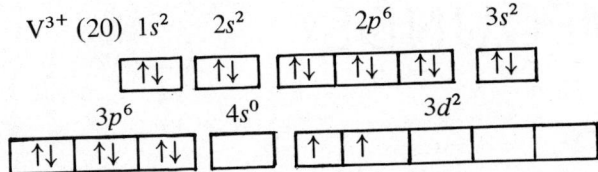

Question 65

Show the electron configuration for the first excited state of the O^{2-} ion.

Answer 65

O^{2-} (10) $1s^2$ $2s^2$ $2p^5$ $3s^1$

Question 66

Are these atoms and ions paramagnetic or diamagnetic?

a. Mg
b. In
c. Na
d. V
e. W
f. S^{2-}
g. V^{3+}
h. O^{2-}, first excited state

Answer 66

a. diamagnetic
b. paramagnetic
c. paramagnetic
d. paramagnetic
e. paramagnetic
f. diamagnetic
g. paramagnetic
h. paramagnetic

PART 4

COMPOUNDS

IONIC COMPOUNDS

Statement 1

When two elements differing sufficiently in **electron affinity** (attraction for electrons) are brought together, the one with greater affinity pulls away electrons from the other element. The resulting products of this donation of electrons are **negative ions** (from the more electron-attracting element) and **positive ions** (from the less electron-attracting element).

Electron affinities of the **representative elements** (Groups 1A, 2A, 3A, 4A, 5A, 6A, and 7A) increase from left to right across a period and from bottom to top within a group.

Question 1

Which element should have the greater electron affinity, S or Mg?

Answer 1

S

S and Mg are in the same period, with S farthest to the *right* of Mg. Thus S has greater electron affinity.

Question 2

Which element would be more likely to form a positive ion when sodium and chlorine are brought together?

Answer 2

Na

Na, on the extreme left of the periodic chart, has less electron affinity than Cl, which is on the extreme right of the chart.

Statement 2

Electron-dot pictures (or "Lewis" pictures) are often used to depict what happens during ion formation. The electron-dot picture for an element is simply the symbol for the element with dots drawn around it to represent the **valence shell electrons** (those in the highest principal quantum level). Though these dots are usually shown in pairs, their positions are not significant.

Elements from **groups 1A, 2A, and 3A** form **positive ions** by **losing** the number of electrons equal to their group numbers.

Elements from **groups 6A and 7A** form **negative ions** by **gaining** the number of electrons equal to 8 minus the group number.

Question 3

Show the electron-dot picture for

a. the sodium atom

b. the sodium ion

Answer 3

a. Na·

b. Na$^+$

Na is in group 1A, so it has *one* valence electron, which it loses to form a 1+ ion.

Question 4

Show the electron-dot picture for

a. the chlorine atom

b. the typical chlorine ion

Answer 4

a. $:\overset{..}{\underset{..}{Cl}}\cdot$

b. $:\overset{..}{\underset{..}{Cl}}:^-$

Cl is in group 7A, so it has 7 valence electrons. It gains 8−7=1 electron.

Question 5

Show the electron-dot pictures for the ions formed when barium and sulfur are brought together.

Answer 5

Ba^{2+} and $:\!\ddot{\underset{..}{S}}\!:^{2-}$

$\cdot Ba \cdot \rightarrow Ba^{2+}$ (loses two electrons)

$:\!\dot{\underset{\cdot}{S}}\!: \rightarrow :\!\ddot{\underset{..}{S}}\!:^{2-}$ (gains two electrons)

Question 6

Show the electron-dot pictures for the ions formed when aluminum reacts with bromine.

Answer 6

Al^{3+} and $:\!\ddot{\underset{..}{Br}}\!:^{-}$

$\cdot \underset{\cdot}{Al} \cdot \rightarrow Al^{3+}$ (loses three electrons)

$:\!\ddot{\underset{..}{Br}}\!\cdot \rightarrow :\!\ddot{\underset{..}{Br}}\!:$ (gains one electron)

Statement 3

The **representative elements** usually form ions having an electron configuration of the type $ns^2 \, np^6$, where n is the highest principal quantum number of the atom. This $ns^2 \, np^6$ configuration is typical of the **noble gases** in group 8A (or group O, as it is sometimes called).

Question 7

Show the electron configuration for the ions formed when potassium reacts with oxygen.

Answer 7

K^+ (18) $1s^2\ 2s^2\ 2p^6\ 3s^2\ 3p^6$

O^{2-} (10) $1s^2\ 2s^2\ 2p^6$

Question 8

What two noble gases are isoelectronic with (have the same number of electrons as) the ions formed when potassium and oxygen react?

Answer 8

K^+ is like Ar.

O^{2-} is like Ne.

Question 9

Show the electron configurations of the ions formed when magnesium reacts with chlorine, and tell which noble gases have the same configuration.

Answer 9

Mg^{2+} (10) $1s^2\ 2s^2\ 2p^6$; like Ne

Cl^- (18) $1s^2\ 2s^2\ 2p^6\ 3s^2\ 3p^6$; like Ar

Question 10

What noble gas has the same configuration as the ion formed from

a. aluminum?

b. lithium?

Answer 10

a. Ne

b. He (Note that He is the only noble gas with a simple ns^2 configuration.)

Question 11

What noble gas has the same configuration as the ion formed from

a. Rb?

b. Ca?

Answer 11

a. Kr

b. Ar

Question 12

What noble gas has the same configuration as the ion formed from

a. sulfur?

b. bromine?

Answer 12

a. Ar

b. Kr

Question 13

What noble gas has the same configuration as the ion formed from

a. Ba?

b. I?

Answer 13

a. Xe

b. Xe

Statement 4

Positive and negative ions combine in a **crystal lattice** to form an **ionic compound** held together by **electrostatic** (or "electrovalent") bonds.

The overall ionic compound must be neutral; therefore the same number of positive and negative charges must be present in the compound. Hence, if the absolute values of the charges on the two types of ion making up an ionic compound are the **same** (e.g., +1 and −1 or +2 and −2), the **formula** will be of the type MX, where M represents the **cation** (positive ion) and X the **anion** (negative ion).

Question 14

What is the formula for potassium bromide?

Answer 14

KBr

Potassium bromide contains K^+ and Br^- ions.

Question 15

What is the formula for barium oxide?

Answer 15

BaO

Barium oxide contains Ba^{2+} and O^{2-} ions.

Statement 5

When the absolute values of the charges on the cation and anion **differ**, the formula of the ionic compound is of the type $M_x X_m$, where x is the absolute value of the charge on X and m is the absolute value of the charge on M.

Question 16

What is the formula for potassium sulfide?

Answer 16

K_2S

Because there are both K^+ and S^{2-} ions, there must be twice as many K^+ as S^{2-} to give electroneutrality.

Question 17

What is the formula for magnesium chloride?

Answer 17

$MgCl_2$

$(Mg^{2+}$ and Cl^- ions)

Question 18

What is the formula for aluminum oxide?

Answer 18

Al_2O_3

$(Al^{3+}$ and O^{2-} ions)

Question 19

What is the formula for aluminum fluoride?

Answer 19

AlF_3

COVALENT COMPOUNDS

Statement 6

Covalent compounds, which consist of **molecules** rather than ions, are formed between elements differing slightly (or not at all) in electronegativity. In these compounds two elements **share** a pair of electrons to form a covalent bond.

In most compounds this sharing gives each atom in the molecule the same number of electrons in its valence shell as are found in one of the inert gases. This can be shown in an electron-dot representation (**Lewis structure**) of the molecule. It is helpful to designate the electrons (e) from different elements by different symbols, using, for example, solid dots and open dots.

PART 4—COMPOUNDS

Example:

H· + ·C̈l: → H:C̈l: 8e in valence shell of Cl; as in Ar

1e   7e   2e in valence shell of H; as in He

Question 20
Draw the Lewis structure for the F_2 molecule.

Answer 20

:F̈:F̈:

Each F shares one of its 7 electrons so that each has 8 electrons in its valence shell (like Ne).

Question 21
How many covalent bonds are there in the F_2 molecule?

Answer 21
One.

Only one pair of electrons is shared by the two atoms.

Question 22
Draw the Lewis structure for the water molecule (H_2O).

Answer 22

H:Ö:H

Question 23
How many covalent bonds are there in the H_2O molecule?

Answer 23

Two.

Question 24

a. Draw the Lewis structure for ammonia, NH_3.

b. How many covalent bonds hold this molecule together?

Answer 24

a. H : N̈ : H
      ̤
      H

b. Three.

Question 25

a. Draw the Lewis structure for methane, CH_4.

b. How many covalent bonds hold the methane molecule together?

Answer 25

a.   H
   H : C̈ : H
     ̤
     H

b. Four.

Question 26

Assuming that all unshared electrons are paired in the molecule, how many pairs of *unshared* valence electrons (nonbonding electrons) are there in:

a. HCl?

b. H_2O?

c. NH_3?

d. CH_4?

Answer 26

a. 3 pairs.

b. 2 pairs.

c. 1 pair.

d. 0 pair.

Question 27

Consider the polyatomic hydroxide ion, OH^-, to be formed by combining an oxide ion, O^{2-}, and a hydrogen ion, H^+. Show the Lewis structure of the OH^- ion.

Answer 27

$$:\!\overset{..}{\underset{..}{O}}\!:H^-$$

It doesn't matter whether a given electron came originally from O or H; one electron is just like another.

Question 28

Consider the polyatomic ammonium ion, NH_4^+, to be made up of an ammonia molecule and a hydrogen ion. Draw the Lewis structure for the NH_4^+ ion.

Answer 28

$$\begin{array}{c} H \\ \!\!\!\!:: \\ H:\overset{..}{\underset{..}{N}}:H^+ \\ H \end{array}$$

Question 29

Show the Lewis structure for the carbon tetrachloride molecule, CCl_4.

Answer 29

$$\begin{array}{c} :\!\overset{..}{\underset{..}{Cl}}\!: \\ :\!\overset{..}{\underset{..}{Cl}}\!:\overset{..}{\underset{..}{C}}:\overset{..}{\underset{..}{Cl}}\!: \\ :\!\overset{..}{\underset{..}{Cl}}\!: \end{array}$$

Question 30

Draw the Lewis structure for the BCl₃ molecule.

Answer 30

$$:\!\ddot{\underset{..}{Cl}}:B:\ddot{\underset{..}{Cl}}:$$
$$:\underset{..}{Cl}:$$

> Note that B here has only 6e in its outermost shell rather than the usual 8.

Statement 7

A single bond is formed by the sharing of one pair of electrons.

Double and **triple bonds** may be formed by elements sharing 2 or 3 pairs of electrons but still maintaining the noble gas configuration for each partner in the bonds.

Example:

$$\cdot\dot{\underset{.}{N}}: \;+\; :\dot{\underset{.}{N}}\cdot \;\rightarrow\; :N::N: \;\text{ or }\; N \equiv N$$

> 8e in the valence shell of each N

Note that **lines** may be used rather than dots to depict bonds, with one line replacing a pair of dots.

Question 31

What is the Lewis structure for the cyanide ion, CN⁻?

Answer 31

$$:C::N: \;\text{ or }\; C \equiv N^-$$

> We may consider CN⁻ to be formed from a C atom and a N⁻ ion. Compare this structure with that for N₂ in Statement 7.

Question 32

What is the Lewis structure for acetylene, C₂H₂?

Answer 32

H : C :: :: C : H   or   H - C ≡ C - H

Question 33

What is the Lewis structure for ethylene, C_2H_4?

Answer 33

$$H:C::C:H \quad \text{or} \quad \overset{H}{\underset{H}{>}}C=C\overset{H}{\underset{H}{<}}$$

VALENCE-BOND THEORY

Statement 8

The **valence-bond (V-B) theory** gives a more precise picture of covalent bonding. V-B theory proposes that a bond is formed when an **atomic orbital** containing **one unpaired electron** approaches an atomic orbital containing an unpaired electron on another atom. Electrons from both atomic orbitals may occupy **either** atomic orbital; or we may say that the orbitals **overlap**.

Bond formation may be depicted using an **orbital diagram** based on the "central" atom in the molecule.

Example: the water molecule (H_2O)

H (1e) $1s^1$ (one unpaired electron)

O (8e) $1s^2$ $2s^2$ $2p^4$

[↑↓] [↑↓] [↑↓] [↑] [↑]   (two unpaired electrons)

V-B orbital diagram for H_2O:

O 1s   2s   2p

[↑↓] [↑↓] [↑↓] [↑↓] [↑↓]   Each H supplies one electron for pairing with an electron in a half-
         H   H   full orbital in O.

Question 34

Show the V-B orbital diagram for NH_3.

Answer 34

for N atom:

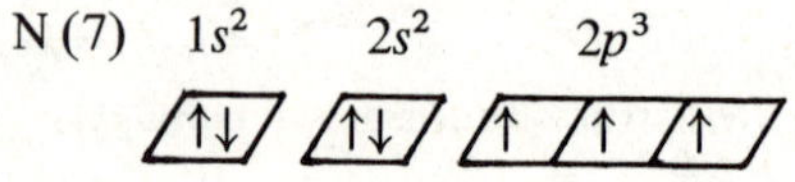

$N(7)$ $1s^2$ $2s^2$ $2p^3$

[↑↓] [↑↓] [↑] [↑] [↑]

for NH_3 molecule:

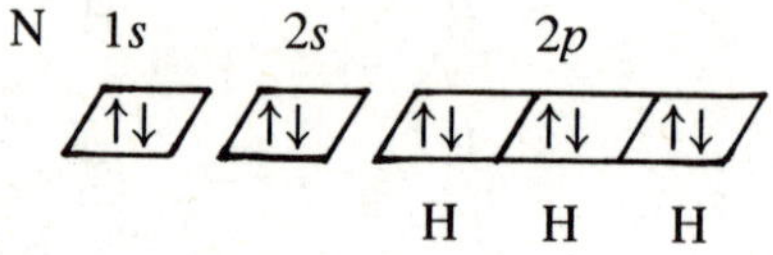

N   1s   2s   2p

[↑↓] [↑↓] [↑↓] [↑↓] [↑↓]
              H   H   H

Question 35

Show the V-B orbital diagram for H_2S.

Answer 35

for S atom:

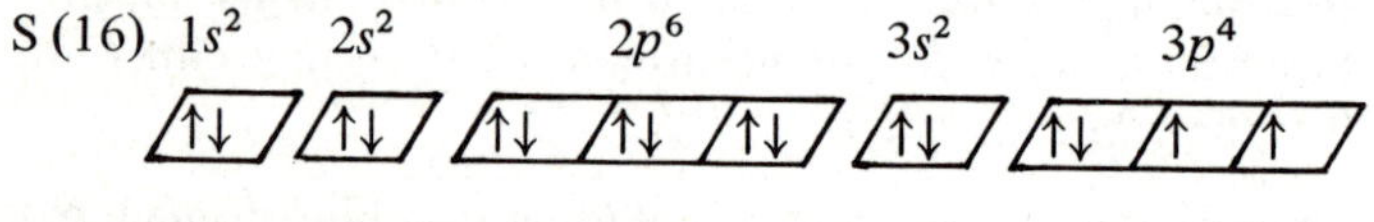

$S(16)$ $1s^2$ $2s^2$ $2p^6$ $3s^2$ $3p^4$

[↑↓] [↑↓] [↑↓] [↑↓] [↑↓] [↑↓] [↑↓] [↑] [↑]

for the H_2S molecule:

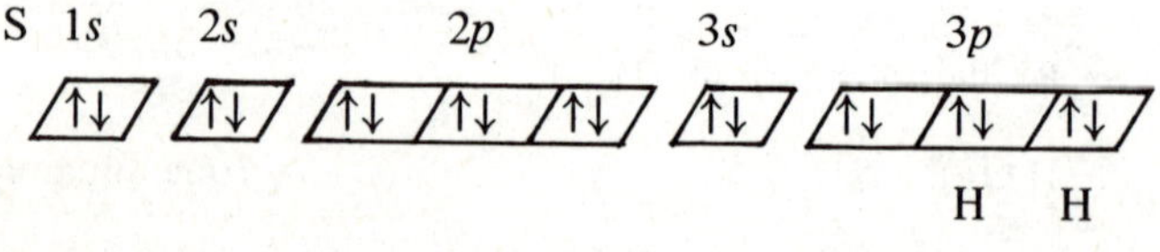

S 1s   2s   2p   3s   3p

[↑↓] [↑↓] [↑↓] [↑↓] [↑↓] [↑↓] [↑↓] [↑↓] [↑↓]
                                    H   H

Question 36

Show the V-B orbital diagram for the HF molecule.

Answer 36

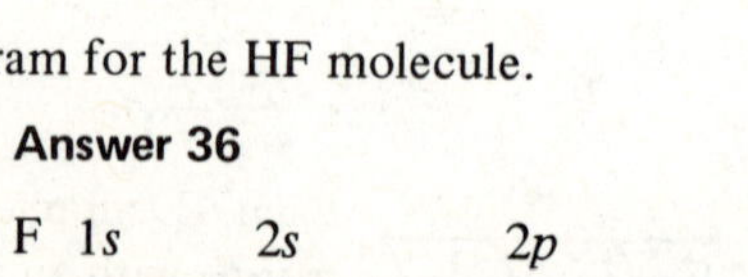

F 1s   2s   2p

[↑↓] [↑↓] [↑↓] [↑↓] [↑↓]
                     H

Statement 9

Hybridization of orbitals is a concept required by V-B theory to account for the fact that in some molecules more bonds are formed than would be expected from the predicted electron configurations of the atoms involved. This approach suggests that some of the atomic orbitals on the central atom are changed during the bonding process to give a new set of hybrid atomic orbitals, all exactly alike, suitable for bonding.

Example: the BH_3 molecule

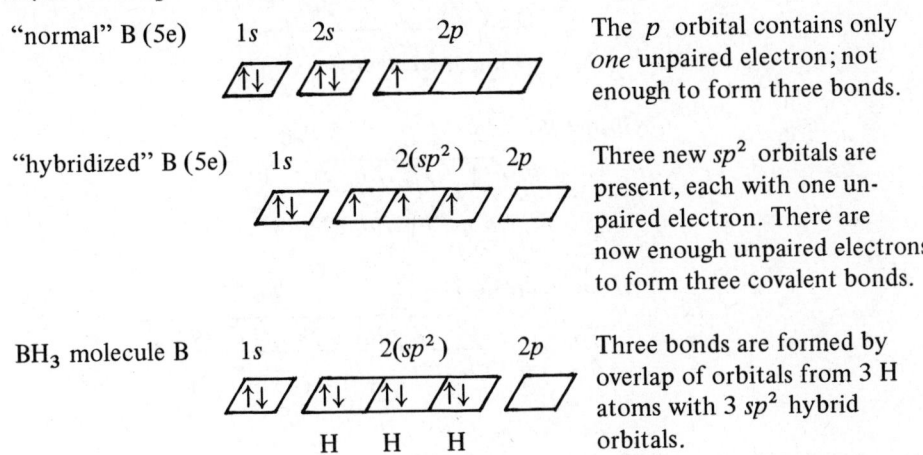

Thus B may form three bonds although "normal" B has only one unpaired electron.

Question 37

Show an orbital diagram for the hybridization that would account for the formation of BeF_2, and name the hybrid orbitals produced.

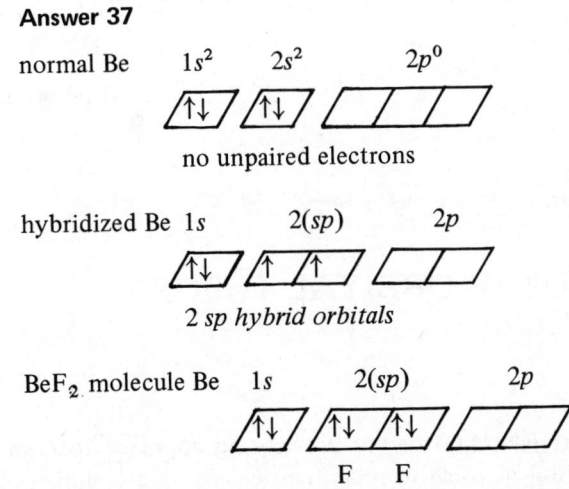

Question 38

Show an orbital diagram for the hybridization in CH_4, and name the hybrid orbitals produced.

Answer 38

normal C $1s^2$ $2s^2$ $2p^2$

$\boxed{\uparrow\downarrow}\,\boxed{\uparrow\downarrow}\,\boxed{\uparrow}\,\boxed{\uparrow}\,\boxed{}$

2 unpaired electrons

hybridized C $1s^2$ $2(sp^3)$

$\boxed{\uparrow\downarrow}\,\boxed{\uparrow}\,\boxed{\uparrow}\,\boxed{\uparrow}\,\boxed{\uparrow}$

4 sp^3 hybrid orbitals

CH_4 molecule C $1s^2$ $2(sp^3)$

$\boxed{\uparrow\downarrow}\,\boxed{\uparrow\downarrow}\,\boxed{\uparrow\downarrow}\,\boxed{\uparrow\downarrow}\,\boxed{\uparrow\downarrow}$

       H H H H

Question 39

Show an orbital diagram for the hybridization in CHF_3, and name the hybridized orbitals produced.

Answer 39

CHF_3 molecule C $1s$ $2(sp^3)$

$\boxed{\uparrow\downarrow}\,\boxed{\uparrow\downarrow}\,\boxed{\uparrow\downarrow}\,\boxed{\uparrow\downarrow}\,\boxed{\uparrow\downarrow}$

       H F F F

sp^3 hybrid orbitals

MOLECULAR ORBITAL THEORY

Statement 10

Molecular orbital (M-O) theory provides an approach to covalent bonding that is even more successful in some respects, particularly in accounting for magnetic properties of molecules, than is the V-B theory.

M-O theory suggests that when two atoms approach each other, a new set of molecular orbitals is formed. These are then filled in an **aufbau process**, using valence electrons from the bonding atoms.

Molecular orbitals may be either **sigma**, σ (symmetrical about an axis joining the bonded atoms), or **pi**, π (symmetrical above and below a plane passing through the bonded atoms).

These σ and π orbitals may be either **bonding orbitals** (σ or π) or **antibonding orbitals** (σ^* or π^*). Electrons placed in a bonding orbital serve to **destabilize** it. Each filled antibonding orbital **cancels** the stabilization produced by a similar filled bonding orbital.

An **electron pair bond** exists when there is a **net** of **two bonding electrons** in the molecule.

Summary of Procedure

1. Add up the valence electrons from each of the two atoms involved.
2. Fill molecular orbitals until valence electrons are used up. Place each electron in the lowest vacant energy level, observing Hund's rule and using the following energy ordering:

$$\sigma_s \quad \sigma_s^* \quad \pi_{p_x} \quad \pi_{p_y} \quad \sigma_{p_z} \quad \pi^*_{p_x} \quad \pi^*_{p_y} \quad \sigma^*_{p_z}$$

increasing energy →

Note: Some texts place the σ_{p_z} level **below** the π_{p_x}–π_{p_y} level, which appears to be correct for O_2 and F_2, but not for lower elements. In practice, correct predictions regarding numbers and types of bonds will be made using the energy ordering shown above.

Question 40

a. Show the M-O diagram for the H_2 molecule.

b. How many electron-pair bonds hold the molecule together?

Answer 40

a. $\begin{array}{c} \text{H (1 valence e)} \\ \text{H (1 valence e)} \end{array} > \begin{array}{c} \sigma_s^2 \\ \boxed{\uparrow\downarrow} \end{array}$

b. one, a σ_s bond

Question 41

Using the M-O diagram approach, predict the type of bonding in the B_2 molecule.

Answer 41

One π electron-pair bond.

M-O diagram:

B (3 valence e)
B (3 valence e) $>$ ϕ_s^2 ϕ_s^{*2} $\pi_{p_x}^{\,1}$ $\pi_{p_y}^{\,1}$

[↑↓] [↑↓] [↑] [↑]

These cancel. → 2 net electrons in *bonding* orbitals.

Question 42

How many unpaired electrons are there in the B_2 molecule?

Answer 42

Two

Examining the M-O diagram, we see that the π_{p_x} and π_{p_y} orbitals each contain one electron.

Question 43

a. Show the M-O diagram for the N_2 molecule.

b. Predict the type of bonding in the N_2 molecule.

Answer 43

a. N (5 valence e)
   N (5 valence e) $>$ ϕ_s^2 ϕ_s^{*2} $\pi_{p_x}^{\,2}$ $\pi_{p_y}^{\,2}$ $\sigma_{p_z}^{\,2}$

[↑↓] [↑↓] [↑↓] [↑↓] [↑↓]

b. 3 e-pair bonds, 2 π and one σ.

PART 4—COMPOUNDS

Question 44

a. Show the M-O diagram for the O_2 molecule.

b. Should it be paramagnetic?

Answer 44

a. O (6 valence e) > O (6 valence e)

$\phi_s^2 \quad \phi_s^{*2} \quad \pi_{p_x}^2 \quad \pi_{p_y}^2 \quad \sigma_{p_z}^2 \quad \pi_{p_x}^{*1} \quad \pi_{p_y}^{*1}$

| ↑↓ | ↑↓ | ↑↓ | ↑↓ | ↑↓ | ↑ | ↑ |

b. Yes; there are two *unpaired* electrons.

Question 45

Predict the type of bonding in the CO molecule and the number of unpaired electrons.

Answer 45

3 e-pair bonds, 0 unpaired e

M-O diagram:

$\phi_s^2 \quad \phi_s^{*2} \quad \pi_{p_x}^2 \quad \pi_{p_y}^2 \quad \sigma_{p_z}^2$

C (4 valence e) > O (6 valence e)

| ↑↓ | ↑↓ | ↑↓ | ↑↓ | ↑↓ |

3 net bonds

Question 46

Show the M-O diagram for the NO molecule, and decide on the type of bonding and number of unpaired electrons in the molecule.

Answer 46

$\phi_s^2 \quad \phi_s^{*2} \quad \pi_{p_x}^2 \quad \pi_{p_y}^2 \quad \sigma_{p_z}^2 \quad \pi_{p_x}^{*1} \quad \pi_{p_y}^{*0}$

N (5 valence e) > O (6 valence e)

| ↑↓ | ↑↓ | ↑↓ | ↑↓ | ↑↓ | ↑ | |

half cancelled by $\pi_{p_x}^*$ electron

$2\frac{1}{2}$ e-pair bonds.

1 unpaired electron.

Question 47

Predict the type of bonding in the CN⁻ ion.

Answer 47

3 e-pair bonds, 2π and 1σ (like the isoelectronic N_2 molecule).

GEOMETRY OF MOLECULES

Statement 11

The chemist is often interested in knowing something about the **geometry** of a molecule, which describes the spatial arrangement of the atoms with respect to one another.

In the V-B approach to predicting geometries, the directional properties of the orbitals involved in bonding are taken into account. The types of angles they form with one another can be calculated by V-B theory. **Prediction for a specific molecule** is accomplished in the following manner:

a. Inspect the molecular formula to see how many atoms, a, are bonded to the central atom, c.

b. Determine the orbital condition of c (i.e., whether hybridization of orbitals from c are used) (see Statement 9 in this chapter).

c. Relate the molecular formula type and orbital condition to the geometry. A listing for some common types is given.

Molecular Formula Type	Orbital Condition	Geometry
ca_2	sp	linear
ca_2	p^2	angular
ca_3	sp^2	trigonal planar
ca_3	p^3	trigonal pyramidal
ca_4	sp^3	tetrahedral
ca_4	dsp^2	square planar
ca_5	dsp^3	trigonal bipyramidal
ca_6	d^2sp^3	octahedral

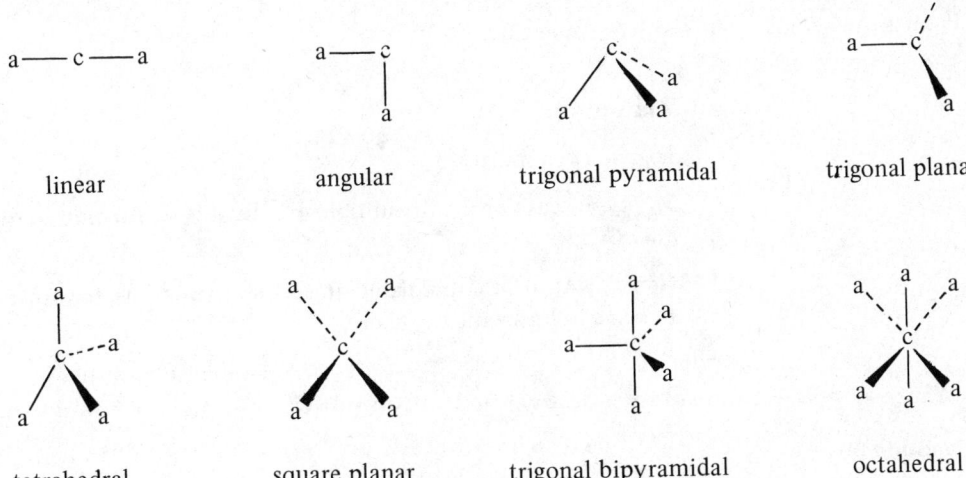

linear angular trigonal pyramidal trigonal planar

tetrahedral square planar trigonal bipyramidal octahedral

Note: Solid bonds (———) are in the plane of the paper.
Dashed bonds (- - - -) project behind the plane of the paper.
Tapered bonds (▬▬) project in front of the plane of the paper.

Question 48

Predict the geometry of the H$_2$O molecule.

Answer 48

angular

MF type—ca$_2$
orbital condition predicted to be:

O 1s^2 2s^2 2p^4

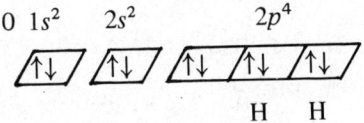

H   H

p^2 (two orbitals used in bonding are p orbitals)

Therefore, *angular* geometry.

Note: We would further predict a 90° angle for H-O-H. It is actually 104°, so some hybridization of orbitals is suspected.

Question 49

Predict the geometry of the NH_3 molecule.

Answer 49

trigonal pyramidal

MF type—ca_3; orbital condition—p^3; therefore, *trigonal pyramidal*.

Note: Again, the predicted angle turns out to be too small, so some hybridization is likely.

Question 50

Predict the geometry of the BF_3 molecule.

Answer 50

tirgonal planar

Question 51

Predict the geometry of the SF_6 molecule.

Answer 51

octahedral

Question 52

The $PtCl_4^{2-}$ ion is square planar. What sort of hybridization must Pt have undergone?

Answer 52

dsp^2

MF type—ca_4
geometry—square planar

Therefore, dsp^2 hybridization.

Question 53

The CO_2 molecule is linear. What sort of orbitals is C using to bond with oxygen?

Answer 53

sp hybrid orbitals

Statement 12

The **electron-pair repulsion method** (EPR) is a simpler approach to predicting geometries of molecules. We assume that all electron-pairs in the valence shell take up **lone-pair** positions; that is, because they **repel** one another (being of like charge) and the electron-pairs used for bonding, they occupy positions that will get them as far away as possible from other electron pairs.

We may represent a molecule, then, as cb_xL_y, where c is the central atom, b the bonding-pairs (with their atoms), and L the lone-pairs.

Procedure:

a. Write the orbital diagram for the molecule.

b. Count the number of "lone-pairs" remaining in the outermost shell of the central atom.

c. Arrange the bonding-pairs and lone-pairs about the central atom to give **minimum repulsion geometry** (minimum repulsion among electron-pairs). The order of repulsion is:

$$b\text{-}b < L\text{-}b < L\text{-}L$$
increasing repulsion →

d. Name the geometry of the molecule, based on the spatial relationship between the central atoms and the atoms bonded to it.

total e-pairs	minimum repulsion geometry
2	linear
3	trigonal planar
4	tetrahedral
5	trigonal bipyramidal
6	octahedral

Question 54

Predict the geometry of the H₂O molecule, using the EPR method.

Answer 54

angular

orbital diagram:

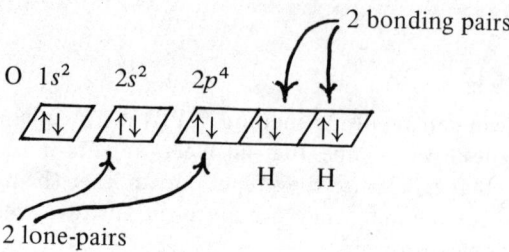

A total of four pairs of electrons in the valence shell must be placed around the O atom. The distance separating all four will be greatest with a tetrahedral arrangement:

minimum repulsion geometry

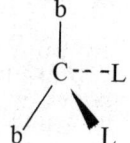

The geometry of the *molecule* is based on the arrangement of bonded atoms only, and the b-c-b geometry is *angular* (109.5°).

Note: Actual H-O-H angle is 104.5° rather than the tetrahedral angle of 109.5°, because of the L-b repulsion pushing the O-H bonding pairs closer together.

Question 55

For the NH₃ molecule, predict:

a. the "minimum repulsion geometry" for electron-pairs.
b. the geometry of the molecule.

Answer 55

a. tetrahedral
b. trigonal pyramidal

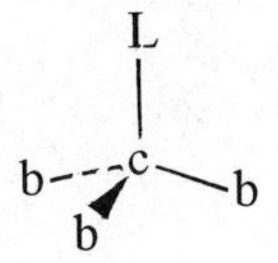

| tetrahedral | trigonal pyramidal |

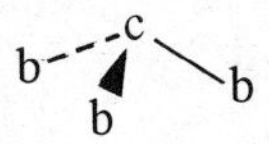

Question 56

Should the H-N-H bond angle in NH_3 be *exactly* 109.5° (the tetrahedral angle)?

Answer 56

No.

H-N-H angles should be *less* than the tetrahedral value because of the *L-b* repulsion pushing N-H bonding-pairs closer together. (It is actually 107.3°.)

Question 57

Predict the geometry of the CH_4 molecule.

Answer 57

tetrahedral

4 *b-b* e-pairs at maximum angles (109.5°) from one another. (All repulsions here are *b-b* type, so the H-C-H angle is exactly 109.5°.)

Question 58

Predict the geometry of the PF_5 molecule.

Answer 58

trigonal bipyramidal

Five *b-b* e-pairs as far apart as possible.

Question 59

Predict the geometry of the SF$_4$ molecule, using the EPR method.

Answer 59

Orbital diagram:

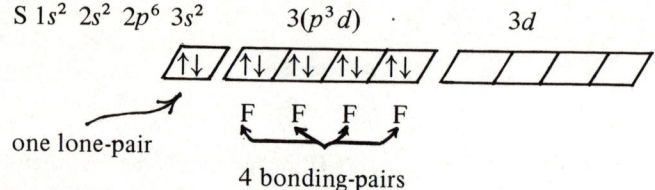

There is a total of five pairs to arrange around the central S atom, and two possible arrangements (both trigonal bi-pyramidal):

Greatest repulsion is between *closest* e-pairs (90° apart); L-b repulsion is greater than b-b repulsion. The first arrangement minimizes repulsion—2 L-b repulsions (↔↔) rather than 3. Ignoring the lone-pair gives the molecular geometry shown above for SF$_4$.

Question 60

Predict the geometry of the ClF$_3$ molecule.

Answer 60

T-shaped   F—Cl—F
                |
                F

From the orbital diagram we see that the central Cl atom must have undergone p^2d hybridization in bonding with 3 F atoms, giving 2 L and 3 b.

Possibilities:

```
      L                    L                    b
      |                    |                    |
      |..L                 |..b                 |..L
   b—C                  b—C                  b—C
      |`b                  |`b                  |`L
      |                    |                    |
      b                    L                    b
```

repulsions 1 L-L   6 L-b   4 L-b
          3 L-b   minimum
                                               repulsion

Ignoring 2 L's, the result is a T-shaped molecule.

Statement 13

A **polar bond** is formed in a molecule whenever the two atoms sharing electrons differ in **electronegativity**, the attraction of an atom in a molecule for electrons. Electronegativity, like electron affinity, increases from left to right across a period and from bottom to top in a group on the periodic table. A polar bond is symbolized by placing δ^- on the more negative end and δ^+ on the more positive end of the bond:

$$\overset{\delta^-}{X}\!\!\longrightarrow\!\!\overset{\delta^+}{Y}$$

In addition, a **polar molecule** will result if more negative charge (or electron density) is concentrated on one side of the molecule. Such polarity has a profound effect on the chemistry of the molecule. The presence of one or more polar bonds in a molecule does not guarantee that the molecule itself will be polar. The polarity or nonpolarity of a molecule depends on three factors:

a. **bond polarities**—Do differences in electronegativities of bonded atoms produce charge separation in the molecule?

b. **molecular geometry**—Does the shape of the molecule result in the cancellation by a polar bond on one side of the effect of an identical bond on the other side?

c. **lone-pair electron positions**—Are any lone-pairs located in such a way that they balance one another?

The important question, in summary, is always this: Is the electron density on one side of the molecule **exactly balanced** by the electron density on the other side? If the answer is no, the molecule is polar.

Question 61

Which of these molecules contain polar bonds?

a. N_2

b. NO

c. $CHCl_3$

d. Cl_2

Answer 61

(b) and (c) only.

In the NO molecule, O is more electronegative than N. In $CHCl_3$, Cl is more electronegative than H. In Cl_2 and N_2 both atoms involved in the bonds have the same electronegativity.

Question 62

Show with δ^+ and δ^- symbols the polarities of these bonds:

a. CO_2 (linear geometry)

b. HF

Answer 62

a. $\overset{\delta^- \;\; \delta^+ \;\; \delta^-}{O=C=O}$

b. $\overset{\delta^+ \;\; \delta^-}{H-F}$

Question 63

Which of these molecules are polar?

a. CO_2

b. H_2O

c. HF

Answer 63

b. H_2O is angular, with 2 lone-pairs on one side of molecule.

c. HF is linear, with F more electronegative than H.

a. CO_2 is nonpolar. It is linear: the O on one side is exactly balanced by the O on the opposite side.

Question 64

Which of these molecules would you expect to be nonpolar?

a. CH_4

b. SF_4

c. NH_3

Answer 64

a. CH_4 is tetrahedral, with the same kind of atom at every point, giving exact balance of electrons about the central C atom.

b. SF_4 has F's on one side of the molecule but a lone-pair on the other (see Answer 59) and therefore is polar.

c. NH_3 is trigonal pyramidal and polar, with 3 H's on one side and a lone-pair on the other.

Question 65

SiH_4 has been found experimentally to be nonpolar. Is its geometry like that of CH_4 (tetrahedral) or like that of SF_4?

Answer 65

Like CH_4: tetrahedral.

If its geometry were like that of SF_4, SiH_4 would be polar even though it has no lone-pair electrons. The tetrahedral arrangement places the 4 H's in such a way that there is exact balance of electron density in the molecule.

NAMING INORGANIC COMPOUNDS

Statement 14

To name an **ionic compound**:

(a) First write the name of the positive ion. This will be the usual name of the metal forming the ion, except in those cases in which a polyatomic positive ion is involved (see Table 4-1).

(b) When a metal forms more than one ion, place the Roman numeral corresponding to the positive charge on the metal ion in parentheses after the name of the metal.

(c) Write the name of the negative ion last. This name will be formed in most cases by dropping the usual ending of the name of the element and replacing it with "ide." When a polyatomic negative ion is involved, its name will be written as is (see Table 4-1).

Question 66

Name these compounds:

a. NaBr

b. Rb_2O

c. MgS

Answer 66

a. sodium bromide

b. rubidium oxide

c. magnesium sulfide

TABLE 4-1 POLYATOMIC IONS

Ion	Name	Ion	Name
$C_2H_3O_2^-$	acetate	$Cr_2O_7^{2-}$	dichromate
NH_4^+	ammonium	OH^-	hydroxide
BiO^+	bismuthyl	ClO^-	hypochlorite
CO_3^{2-}	carbonate	NO_3^-	nitrate
ClO_3^-	chlorate	ClO_4^-	perchlorate
ClO_2^-	chlorite	MnO_4^-	permanganate
CrO_4^{2-}	chromate	PO_4^{3-}	phosphate
CN^-	cyanide	SO_4^{2-}	sulfate
		SO_3^{2-}	sulfite

Question 67

Name these compounds:

a. NH_4I

b. Li_2SO_4

c. $KMnO_4$

Answer 67

a. ammonium iodide

b. lithium sulfate

c. potassium permanganate

Question 68

Name these compounds:

a. NH_4ClO_4

b. $BiOCl$

c. $Ca(ClO)_2$

Answer 68

a. ammonium perchlorate

b. bismuthyl chloride

c. calcium hypochlorite

Question 69

Name these compounds:

a. Ag_2CO_3

b. $K_2Cr_2O_7$

c. $NaCN$

Answer 69

a. silver carbonate

b. potassium dichromate

c. sodium cyanide

Question 70

Name these compounds:

a. $FeCl_2$

b. $FeCl_3$

c. $Fe(OH)_3$

Answer 70

a. iron (II) chloride (often called ferrous chloride)

b. iron (III) chloride (often called ferric chloride)

c. iron (III) hydroxide (or ferric hydroxide)

Statement 15

Covalent binary compounds are named in the following manner:

a. Write the name of the less electronegative element first.

b. Write the name of the more electronegative element last, changing its ending to "-ide."

c. When the two elements involved may form more than one compound with each other, use Greek prefixes—di-, tri-, tetra-, penta-, hexa-, and so forth—to indicate how many atoms of each are in the molecule.

Question 71

Name these compounds:

a. HCl

b. SO_2

c. SO_3

Answer 71

a. hydrogen chloride

b. sulfur dioxide

c. sulfur trioxide

Question 72

Name these compounds:

a. NO_2

b. N_2O_4

Answer 72

a. nitrogen dioxide

b. dinitrogen tetroxide

Statement 16

An important group of covalent compounds having the general formula H_aY dissolve in water to form H^+ ions and Y^{a-} ions. These are called **acids** and are not customarily named according to the rules used for other covalent compounds.

Procedure for Naming Acids

a. For **ternary acids**, name the Y portion by writing the name of the Y^{a-} ion (see Table 4-1), making the following changes in endings from ion name to acid name: change "-ate" to "-ic"; change "-ite" to "-ous"; then write the word "acid."

b. For **binary acids**, write the prefix "hydro-" followed by the name of the Y portion, changing the usual ending to "-ic"; then write the word "acid."

Question 73

Name these compounds:

a. H_2SO_3

b. H_2SO_4

Answer 73

a. sulfurous acid

b. sulfuric acid

Question 74

Name these compounds:

a. HCl

b. HI

Answer 74

a. hydrochloric acid

b. hydriodic acid

Note: These compounds when *pure* (as opposed to *dissolved in water*) could be called "hydrogen chloride" and "hydrogen iodide" respectively. Notice, too, that the "o" is omitted between the "r" and "i" in hydriodic acid to facilitate pronunciation.

Question 75

Name these compounds:

a. H_2CrO_4

b. HNO_3

Answer 75

a. chromic acid

b. nitric acid

Question 76

Name these compounds:

a. $HClO_4$

b. $HClO_2$

c. $HClO$

Answer 76

a. perchloric acid

b. chlorous acid

c. hypochlorous acid

TESTING YOUR MASTERY—Part 4 Compounds

Question 77

Name these compounds:

a. $CaCl_2$

b. MgO

c. Cs_2S

> ### Answer 77
>
> a. calcium chloride
>
> b. magnesium oxide
>
> c. cesium sulfide

Question 78

Name these compounds:

a. $(NH_4)_2 SO_4$

b. $Fe(NO_3)_3$

c. K_2CrO_4

> ### Answer 78
>
> a. ammonium sulfate
>
> b. iron (III) nitrate (often called ferric nitrate)
>
> c. potassium chromate

Question 79

Name these compounds:

a. $Al(OH)_3$

b. LiH

c. NO_2

Answer 79

a. aluminum hydroxide

b. lithium hydride

c. nitrogen dioxide

Question 80

Name these compounds:

a. HNO_2

b. $HClO_3$

c. HBr

Answer 80

a. nitrous acid

b. chloric acid

c. hydrobromic acid (or, when not in solution, "hydrogen bromide")

Question 81

Arrange these elements in order of decreasing electron affinity: B O Al

Answer 81

$O > B > Al$

Question 82

Predict for the following pairs of elements those that would be more likely to accept electrons from the other when the pair is brought together.

a. Ba and Br

b. Se and Sr

Answer 82

a. Br

b. Se

Question 83

Show the electron-dot pictures for the ions expected to form when calcium metal and iodine crystals are brought together.

Answer 83

$$\cdot \text{Ca} \cdot \quad \rightarrow \quad \text{Ca}^{2+}$$

$$:\!\ddot{\underset{..}{\text{I}}}\!: \quad \rightarrow \quad :\!\ddot{\underset{..}{\text{I}}}\!:^{-}$$

Question 84

Show the electron-dot pictures for the ions formed when aluminum reacts with oxygen.

Answer 84

$$\cdot \dot{\text{Al}} \cdot \quad \rightarrow \quad \text{Al}^{3+}$$

$$\cdot \ddot{\underset{..}{\text{O}}} \cdot \quad \rightarrow \quad :\!\ddot{\underset{..}{\text{O}}}\!:^{2-}$$

Question 85

a. Show the electron configurations for the ions formed when calcium and iodine react.

b. What noble gases have the same configurations as these ions?

Answer 85

a. $\text{Ca}^{2+}(18)\ 1s^2\ 2s^2\ 2p^6\ 3s^2\ 3p^6$

$\text{I}^{-}(54)\ 1s^2\ 2s^2\ 2p^6\ 3s^2\ 3p^6\ 4s^2\ 3d^{10}\ 4p^6\ 5s^2\ 4d^{10}\ 5p^6$

b. Ar (18)

Xe (54)

Question 86

a. Show the electron configurations for the typical ions formed by gallium and selenium.

b. What noble gases have the same configurations as these ions?

Answer 86

a. Ga^{3+} (28) $1s^2\ 2s^2\ 2p^6\ 3s^2\ 3p^6\ 4s^0\ 3d^{10}\ 3p^0$

Se^{2-} (36) $1s^2\ 2s^2\ 2p^6\ 3s^2\ 3p^6\ 4s^2\ 3d^{10}\ 3p^6$

b. No noble gas is isoelectronic with Ga^{3+} (28).

Kr (36) is isoelectronic with Se^{2-}.

Question 87

What is the formula for the compound formed when strontium and fluorine react?

Answer 87

SrF_2

Question 88

Give the formula for:

a. aluminum sulfide

b. rubidium oxide

Answer 88

a. Al_2S_3

b. Rb_2O

Question 89

What is the formula for:

a. calcium hydride?

b. lithium selenide?

Answer 89

a. CaH_2

b. Li_2Se

Question 90

Give the formula for:

a. perchloric acid

b. sodium sulfite

> **Answer 90**
>
> a. $HClO_4$
>
> b. Na_2SO_3

Question 91

a. Draw the electron-dot picture of the Cl_2 molecule.

b. How many covalent bonds hold this molecule together?

> **Answer 91**
>
> a. $:\ddot{\underset{..}{Cl}}:\overset{\circ\circ}{\underset{\circ\circ}{Cl}}:$
>
> b. one covalent bond

Question 92

a. Draw the Lewis structure of the hydrogen sulfide molecule.

b. How many covalent bonds does it contain?

> **Answer 92**
>
> a. $H:\ddot{\underset{..}{S}}:H$
>
> b. two covalent bonds

Question 93

a. Show the electron-dot picture of the silane (SiH_4) molecule.

b. How many covalent bonds hold it together?

Answer 93

a. 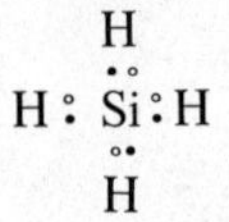

b. four covalent bonds

Question 94

How many pairs of *unshared* valence electrons (nonbonding electrons) are there in:

a. Cl_2?

b. H_2S?

c. SiH_4?

d. $CHBr_3$?

Answer 94

a. 6 pairs

b. 2 pairs

c. 0 pairs

d. 9 pairs

Question 95

Show the Lewis structure of these ions:

a. SH^-

b. NO_3^-

Answer 95

a. 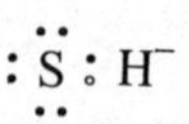

b. 

Question 96

a. Show the Lewis structure for the SO molecule.

b. How many covalent bonds does it contain?

c. How many unshared pairs of electrons does it contain?

Answer 96

a. $\ddot{\underset{..}{S}} :: \overset{\circ\circ}{\underset{\circ\circ}{O}}$

b. two covalent bonds

c. 4 pairs of unshared electrons

Question 97

a. Show the Lewis structure for the HCN molecule.

b. How many covalent bonds does it contain?

c. How many unshared pairs of electrons does it contain?

Answer 97

a. $H \overset{x}{\cdot} C :: N :$

b. 4 covalent bonds, 3 between C and N (triple bond) and one (single bond) between H and C.

c. 1 pair of unshared electrons

Question 98

a. Show the V-B diagram for the HCl molecule.

b. What hybridization is proposed to account for the bonding?

c. How many covalent bonds are formed?

Answer 98

a. for the HCl molecule:

Cl 1s	2s	2p			3s	3p		
↑↓	↑↓	↑↓	↑↓	↑↓	↑↓	↑↓	↑↓	↑↓

H

b. No hybridization required.

c. One covalent bond.

Question 99

a. Show the V-B diagram for the PH$_3$ molecule.

b. What hybridization must be involved in bonding?

c. How many covalent bonds are formed?

Answer 99

a. for the PH$_3$ molecule:

P 1s   2s   2p   3s   3p

[↑↓] [↑↓] [↑↓|↑↓|↑↓] [↑↓] [↑↓|↑↓|↑↓]

                                                                H H H

b. No hybridization required.

c. Three covalent bonds.

Question 100

a. Show the V-B diagram for the BF$_3$ molecule.

b. What hybridization must be proposed to account for the bonding?

c. How many covalent bonds are formed?

Answer 100

a. BF$_3$ molecule:

B   1s   2(sp^2)   2p

[↑↓] [↑↓|↑↓|↑↓] [ ]

           F F F

b. sp^2 hybridization required.

c. Three covalent bonds.

Question 101

a. Show the V-B diagram for the CH$_2$Br$_2$ molecule.

b. What hybridization must be proposed to account for the bonding?

c. How many covalent bonds are formed?

Answer 101

a. CH_2Br_2 molecule:

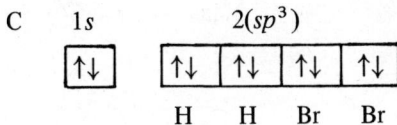

b. sp^3 hybridization required.

c. Four covalent bonds.

Question 102

a. Show the molecular orbital diagram for the SO molecule.

b. How many electron-pair bonds hold the molecule together?

c. Is the molecule paramagnetic?

Answer 102

a. S (6 valence e) >
   O (6 valence e)

$\phi_s^2 \quad \phi_s^{*2} \quad \pi_{p_x}^2 \pi_{p_y}^2 \quad \sigma_{p_z}^2 \quad \pi_{p_x}^{*1} \pi_{p_y}^{*1}$

| ↑↓ | ↑↓ | ↑↓ | ↑↓ | ↑↓ | ↑ | ↑ |

b. Two e-pair bonds.

c. Yes; there are two unpaired electrons.

Question 103

a. Show the molecular orbital diagram for the S_2^{2-} ion.

b. Predict the number of bonds holding the ion together.

c. Should the ion be paramagnetic?

Answer 103

a. $\left.\begin{matrix} S^- \text{ (7 valence e)} \\ S^- \text{ (7 valence e)} \end{matrix}\right\rangle$

$\phi_s^{\ 2}$	$\phi_s^{\ *2}$	$\pi_{p_x}^{\ 2}$	$\pi_{p_y}^{\ 2}$	$\sigma_{p_z}^{\ 2}$	$\pi_{p_x}^{\ *2}$	$\pi_{p_y}^{\ *2}$
↑↓	↑↓	↑↓	↑↓	↑↓	↑↓	↑↓

b. one e-pair bond.

c. No; all electrons are paired.

Question 104

Using the molecular orbital theory, predict whether the HHe molecule might exist.

Answer 104

$\left.\begin{matrix} \text{H (1 valence e)} \\ \text{He(2 valence e)} \end{matrix}\right\rangle$

$\sigma_s^{\ 2}$	$\sigma_s^{\ *1}$
↑↓	↑

Yes, the molecule would be held together by a net of $\frac{1}{2}$ e-pair bond.

Question 105

Predict the type of bonding in the BO molecule.

Answer 105

$2\frac{1}{2}$ e-pair bonds; 2π and $\frac{1}{2}$ σ.

Question 106

Predict the geometry of the PH_3 molecule.

Answer 106

trigonal pyramidal

Question 107

a. Predict the geometry of the PCl$_5$ molecule.

b. Would the molecule be polar?

Answer 107

a. trigonal bipyramidal

b. No

Question 108

Predict the geometry of the ICl$_4^-$ ion.

Answer 108

square planar

Question 109

For the H$_3$O$^+$ ion, predict:

a. the "minimum repulsion geometry"

b. the geometry of the ion

Answer 109

a. tetrahedral

b. trigonal pyramidal

Question 110

For the ICl$_5$ molecule, predict:

a. the "minimum repulsion geometry"

b. the geometry of the molecule

c. whether the molecule would be polar

Answer 110

a. octahedral

b. square pyramidal

c. Yes, it would be polar.

Question 111

Predict which of the following molecules should be polar:

a. $BeCl_2$

b. PCl_3

c. SCl_2

Answer 111

a. nonpolar

b. polar

c. polar

PART 5
GRAM ATOMIC WEIGHTS, MOLES AND FORMULAS

Statement 1

The **gram atomic weight** (GAW) of an element is the number of grams corresponding to its atomic weight.

Question 1

What is the GAW of:

a. silver?

b. sodium?

>   **Answer 1**
>
>   a. 107.87 g
>
>   b. 22.99 g

Question 2

How many GAWs of aluminum does one have if he weighs out 40.47 g of aluminum powder?

>   **Answer 2**
>
>   1.500 GAW
>
>   Defining equation:
>
> $$\#a = \#b \times \text{C.F.}$$
>
>   (See Statement 2 in Part 2.)
>
>   OEIOU:
>
> $$\#\text{GAW} = \#\text{g} \times \frac{\text{GAW}}{\text{g}}$$
>
> $$= 40.47 \text{ g} \times \frac{1 \text{ GAW}}{26.98 \text{ g}}$$
>
> $$= 1.500 \text{ GAW}$$

Question 3

How many GAWs of Ca does 53.44 g correspond to?

Answer 3

1.333 GAW

Statement 2

One gram equivalent weight (GEW) of a given substance (element or compound) **will always react with one GEW of another reactant**, no more and no less. The GEW might be considered the **unit of chemical reactivity** (as opposed to a mass unit or number unit).

Historically, oxygen has been used as the standard for determining GEWs of other elements, either directly or indirectly. Mathematically:

$$\#(GEW)_M = \#(GEW)_O$$

where M is the element in question. Because the GEW for oxygen is defined as being exactly 8.000 ... g, the GEW of an element is defined as the number of grams of the element which will react chemically with (or is otherwise chemically equivalent to) 8.000 ... grams of oxygen. Thus:

$$\frac{\#g\,M}{(GEW)_M} = \frac{\#g\,O}{(GEW)_O}$$

Question 4

A tank contains 10.00 g of oxygen. How many GEW of oxygen does it contain?

Answer 4

1.250 GEW

Defining equation:

$$\#a = \#b \times \text{C.F.}$$

OEIOU:

$$\#GEW = \#g \times \frac{GEW}{g}$$

$$= 10.00\,g \times \frac{1\,GEW}{8.00\ldots g}$$

$$= 1.250\,GEW$$

PART 5—GRAM ATOMIC WEIGHTS, MOLES AND FORMULAS 113

Question 5

40.0 g of calcium metal combines with 16.0 g of oxygen. What is the GEW of calcium?

Answer 5

20.0

Defining equation:

$$\frac{\#g\,M}{(GEW)_M} = \frac{\#g\,O}{(GEW)_O}$$

OEIOU:

$$(GEW)_{Ca} = \frac{\#g\,Ca}{\#g\,O} \times (GEW)_O$$

$$= \frac{40.0\,g\,Ca}{16.0\,g\,O} \times 8.00 \ldots g\,O$$

$$= 20.0\,g\,Ca$$

Question 6

A sample of nickel oxide weighs 4.00 g. After being reduced with hydrogen (removing all oxygen from the nickel oxide), the pure nickel residue weighs 3.14 g. What is the GEW of nickel, based on these data?

Answer 6

29.2 g

Defining equation:

$$\frac{\#g\,M}{(GEW)_M} = \frac{\#g\,O}{(GEW)_O}$$

OEIOU:

$$(GEW)_{Ni} = \frac{\#g\,Ni}{\#g\,NiO - \#g\,Ni} \times GEW_O$$

Statement 3

GAW and GEW are related by this equation:

$$GAW = GEW \times n$$

114   PART 5—GRAM ATOMIC WEIGHTS, MOLES AND FORMULAS

The factor n is often called the **combining capacity** of the element. It may have values of $n = 1, 2, 3$, and so forth.

Question 7

The correct GEW of Ni is 29.36 g. If its combining capacity is 2, what is its GAW?

Answer 7

58.72 g

Defining equation:
(also OEIOU):

$$\text{GAW} = \text{GEW} \times n$$
$$= 29.36 \text{ g} \times 2$$
$$= 58.72 \text{ g}$$

Question 8

The GAW of aluminum is 26.98 g. Its combining capacity is 3. What is its GEW?

Answer 8

8.993 g

OEIOU: $\text{GEW} = \text{GAW}/n$

Statement 4

The **Law of Dulong and Petit** can be useful in determining the atomic weight of an element. Stated in equation form, the law is as follows:

$$\text{GAW} \times \text{specific heat} \approx 6 \text{ cal/}°\text{C}$$

where "specific heat" is the number of calories needed to raise the temperature of one gram of the element by one degree Celsius. The sign $\approx$ means "approximately equal to."

Question 9

The specific heat of silver is found to be 0.0558 cal/g °C. What is the approximate GAW of Ag?

Answer 9

107 g

Defining equation: $GAW_{Ag} \times$ sp. ht. ≈ 6 cal/°C

OEIOU: $GAW_{Ag} \approx \dfrac{6 \text{ cal/°C}}{\text{sp. ht.}} = \dfrac{6 \text{ cal/° C}}{0.0558 \text{ cal/g° C}}$

$= 107$ g

Question 10

The exact GEW of silver is found to be 107.87 g. If its approximate GAW is 107 g, what must be its combining capacity?

Answer 10

1

Defining equation: $GAW = GEW \times n$

OEIOU: $n = \dfrac{GAW}{GEW} \approx \dfrac{107 \text{ g}}{107.87 \text{ g}} \approx 1$

Question 11

What is:

a. the exact GAW of Ag?

b. the exact atomic weight of Ag?

Answer 11

a. 107.87 g

b. 107.87 awu

Defining equation: $GAW = GEW \times n$

$= 107.87$ g $\times$ 1

$= 107.87$ g

Question 12

The specific heat of magnesium is 0.246 cal/g °C. Its GEW is 12.15 g. What is the exact atomic weight of magnesium as calculated from these data?

Answer 12

24.30 awu

Statement 5

The **molecular weight** (MW) of a substance is the ratio of the mass of a molecule of the substance to that of an atom of ^{12}C. Molecular weight is calculated by adding up the atomic weights of the atoms composing the molecule.

Question 13

What is the molecular weight of:

a. carbon monoxide?

b. carbon dioxide?

Answer 13

a. 28.0 awu

$$MW_{CO} = AW_C + AW_O$$
$$= 12.0 \text{ awu} + 16.0 \text{ awu}$$
$$= 28.0 \text{ awu}$$

b. 44.0 awu

$$MW_{CO_2} = AW_C + 2AW_O$$
$$= 12.0 \text{ awu} + 2(16.0 \text{ awu})$$
$$= 44.0 \text{ awu}$$

(Values here were arbitrarily rounded off to three significant figures.)

Note: In subsequent chapters MW calculations will be considered superfluous and so will not be shown in OEIOUs.

Question 14

What is the molecular weight of:

a. carbon tetrachloride?

b. sodium chloride?

PART 5—GRAM ATOMIC WEIGHTS, MOLES AND FORMULAS

Answer 14

a. 153.81 awu

$$MW_{CCl_4} = AW_C + 4AW_{Cl}$$

b. NaCl is an *ionic* compound and hence does not have a *molecular* weight, strictly speaking. Nonetheless, chemists often use the term "molecular weight" for *both* ionic and covalent compounds. In this sense, the MW of NaCl is 58.44 awu.

Statement 6

The **gram molecular weight (GMW)** of a substance is the number of **grams** corresponding to its MW. This quantity is often referred to as a **mole** of the substance.

A mole of any substance, either atomic, ionic, or covalent, contains the **same number**, 6.023×10^{23} (called **Avogadro's number**) of **formula units**. Avogadro's number of any kind of particles constitutes a mole of those particles.

The formula unit for a **covalent compound** is its **molecular formula (MF)**, which shows the actual number of atoms of each element present in a molecule of the substance.

The formula unit for an **ionic compound** is its **empirical formula (EF)**, the simplest whole-number ratio of elements present in the compound.

Question 15

What is the GMW to 4 significant figures of:

a. CCl_4?

b. SO_3?

c. H_2SO_4?

Answer 15

a. 153.8 g

b. 80.06 g

c. 98.07 g

Question 16

What is the weight of a mole of:

a. boric acid (H_3BO_3)?

b. potassium bromide?

Answer 16

a. 61.83 g

b. 119.0 g

Question 17

In one mole of CCl_4,

a. how many formula units are there?

b. how many molecules?

Answer 17

a. 6.023×10^{23} formula units

b. 6.023×10^{23} molecules

CCl_4 is a covalent compound.

Question 18

In one mole of KBr,

a. how many formula units are there?

b. how many ions?

Answer 18

a. 6.023×10^{23} formula units

b. 1.205×10^{24} ions

KBr is an *ionic compound*. For *each* formula unit, KBr, there are *two* ions, K^+ and Br^-. Therefore:

$$\# \text{ions} = 2 \times \# \text{formula units}$$
$$= 2 \times (6.023 \times 10^{23})$$
$$= 12.05 \times 10^{23}$$
$$= 1.205 \times 10^{24}$$

Thus, there are two moles of particles (ions) in one mole of KBr.

PART 5—GRAM ATOMIC WEIGHTS, MOLES AND FORMULAS   119

Question 19

How many moles are in:

a. 39.1 g of C_2H_2?

b. 138.7 g of $CaCl_2$?

c. 19.0 g of Cr?

Answer 19

a. 1.50 moles

b. 1.25 moles

c. 0.365 mole

$$\# \text{moles} = \#g \times \frac{\text{mole}}{g}$$

$$= 39.1 \text{ g} \times \frac{1 \text{ mole}}{26.0 \text{ g}}$$

$$= 1.50 \text{ moles}$$

Question 20

How many moles are in:

a. 34.0 g of ammonia (NH_3)?

b. 474 g of potassium permanganate?

Answer 20

a. 2.00 moles of NH_3

b. 3.00 moles of $KMnO_4$

Statement 7

The empirical formula and molecular formula for a covalent compound are related in this way:

$$MF = n \times EF$$

where n = 1, 2, 3, and so forth. Similarly, for the molecular weight and formula weight (weight as calculated from the empirical formula):

$$MW = n \times FW$$

Question 21

The empirical formula of acetylene is CH. Its molecular weight is found to be about 26. What is its molecular formula?

Answer 21

C_2H_2

$$MF = n \times EF$$

OEIOU: $$MF = \frac{MW}{FW} \times EF$$

$$= \frac{26}{13} \times (CH)$$

$$= 2 \times (CH)$$

$$= C_2H_2$$

Note: In "multiplying" the EF, the numbers must appear as *subscripts*.

Question 22

A compound has a molecular weight of about 137. Its empirical formula is PCl_3. What is its molecular formula?

Answer 22

PCl_3

$$MF = n \times EF$$

$$= \frac{MW}{FW} \times EF$$

$$= \frac{137}{137} \times (PCl_3)$$

$$= PCl_3$$

Statement 8

The **percentage composition** of a compound tells us the number of grams (or any other unit of mass or weight) of each element present in 100 g (or other unit) of the compound. Mathematically:

$$\% A = \frac{g\ A}{g\ \text{compound}} \times 100$$

and so forth for other elements present.

Question 23

A 5.00-g sample of a compound is found to contain 1.13 g of phosphorus and 3.87 g of chlorine. What is its percentage composition?

Answer 23

22.6% P

77.4% Cl

$$\%P = \frac{g\ P}{g\ sample} \times 100$$

$$= \frac{1.13\ g}{5.00\ g} \times 100$$

$$= 22.6\%$$

$$\%Cl = \frac{g\ Cl}{g\ sample} \times 100$$

$$= 77.4\%$$

Question 24

A compound weighing 0.1824 g contains 0.0731 g carbon, 0.0121 g hydrogen, and 0.0972 g oxygen. What is its percentage composition?

Answer 24

40.1% C
6.64% H
53.3% O

Statement 9

The **empirical formula** of a compound can be found conveniently from percentage composition data in the following manner:

1. Determine the number of GAWs of each element present in 100 g of the compound:

$$\#GAW_X = \frac{\%X}{GAW_X} \qquad \#GAW_Y = \frac{\%Y}{GAW_Y}$$

2. Find the simplest whole number ratio among the GAWs found to be present.
3. Write the ratio in formula form.

PART 5—GRAM ATOMIC WEIGHTS, MOLES AND FORMULAS

Note that we can use "% X" rather than "g X" in step (1) because "per cent" means "parts per 100" and we are interested in the number of GAWs per 100 g of compound.

Question 25

A compound is found to be 12.7% carbon, 3.2% hydrogen, and 84.1% bromine. What is its empirical formula?

Answer 25

CH_3Br

Step (1)

$$\# GAW_C = \frac{\%C}{GAW_C} = \frac{12.7 \text{ g C}}{12.0 \text{ g C}/GAW_C} = 1.06 \text{ GAW C}$$

$$\# GAW_H = \frac{\%H}{GAW_H} = \frac{3.2 \text{ g H}}{1.01 \text{ g H}/GAW_H} = 3.2 \text{ GAW H}$$

$$\# GAW_{Br} = \frac{\%Br}{GAW_{Br}} = \frac{84.1 \text{ g Br}}{79.9 \text{ g Br}/GAW_{Br}} = 1.05 \text{ GAW Br}$$

Step (2)

$$\text{ratio of C/Br} = \frac{1.06 \text{ GAW}_C}{1.05 \text{ GAW}_{Br}} \approx 1 \frac{GAW_C}{GAW_{Br}}$$

$$\text{H/Br} = \frac{3.2 \text{ GAW}_H}{1.05 \text{ GAW}_{Br}} \approx 3 \frac{GAW_H}{GAW_{Br}}$$

$$\text{Br/Br} = \frac{1.05 \text{ GAW}_{Br}}{1.05 \text{ GAW}_{Br}} = 1$$

Step (3)

The simplest whole-number ratio is 1C:3H:1Br, so the empirical formula is:

$$CH_3Br$$

Question 26

A sample contains 44.2% Fe and 56.8% Cl. What is its empirical formula?

PART 5—GRAM ATOMIC WEIGHTS, MOLES AND FORMULAS

Answer 26

$FeCl_2$

Question 27

A 5.00-g sample of a compound contained 2.73 g carbon, 0.45 g hydrogen, and 1.82 g oxygen. What is its empirical formula?

Answer 27

C_2H_4O

No need to calculate per cent composition on this one. Just start with:

$$\# \text{ GAW}_C = \frac{\# \text{ g C}}{\text{GAW}_C} = \frac{2.73 \text{ g C}}{12.0 \text{ g C/GAW}_C}$$

and so forth for H and O. Then find simplest whole-number ratio.

Question 28

A compound is 54.6% carbon, 9.0% hydrogen, and 36.4% oxygen. Its molecular weight is found to be 88 ± 1. What is its molecular formula?

Answer 28

$C_4H_8O_2$

Find the empirical formula, C_2H_4O, as usual. Then:

$$MF = \frac{MW}{FW} \times EF$$

$$= \frac{88}{44} \times (C_2H_4O)$$

$$= 2 \times (C_2H_4O)$$

$$= C_4H_8O_2$$

TESTING YOUR MASTERY—Part 5
Gram Atomic Weights, Moles, and Formulas

Question 29

What is the GAW of

a. bismuth?

b. zinc?

c. bromine?

> **Answer 29**
>
> a. 208.98 g
>
> b. 65.37 g
>
> c. 79.91 g

Question 30

What is the GMW of:

a. bromine?

b. chloroform ($CHCl_3$)?

c. ethylene (C_2H_4)?

> **Answer 30**
>
> a. 159.82 g
>
> b. 119.38 g
>
> c. 28.34 g

Question 31

What is the molecular weight of ethylene (C_2H_4)?

> **Answer 31**
>
> 28.34 awu

Question 32

What is the weight of a mole of:

a. bromine?

b. sodium bromide?

c. sulfur trioxide?

> **Answer 32**
>
> a. 159.82 g
>
> b. 102.90 g
>
> c. 80.06 g

Question 33

How many moles are contained in:

a. 57.7 g of CH_2Cl_2?

b. 57.7 g of $BaCl_2$?

> **Answer 33**
>
> a. 0.678 mole
>
> b. 0.277 mole
>
> OEIOU: $\# \text{moles} = \# g \times \dfrac{\text{mole}}{g}$
>
> or $\quad \# \text{moles} = \dfrac{\# g}{\text{GMW}}$

Question 34

How many grams of $KMnO_4$ should be weighed out to give 0.100 mole?

> **Answer 34**
>
> 15.8 g
>
> OEIOU: $\# g = \# \text{moles} \times \dfrac{g}{\text{mole}}$

Question 35

How many grams of aluminum must one weigh out to obtain 0.250 GAWs of the metal?

Answer 35

6.75 g

OEIOU: $\#g = \#GAW \times \dfrac{g}{GAW}$

Question 36

How many formula units and how many ions are in:

a. one mole of LiI?

b. one mole of $KMnO_4$?

Answer 36

a. 6.023×10^{23} formula units of LiI
   1.205×10^{24} ions

b. 6.023×10^{23} formula units of $KMnO_4$
   1.205×10^{24} ions

Question 37

How many formula units and how many molecules are in:

a. one mole of NO_2?

b. one mole of $Ca(NO_3)_2$?

Answer 37

a. 6.023×10^{23} formula units of NO_2; 6.023×10^{23} molecules

b. 6.023×10^{23} formula units of $Ca(NO_3)_2$; no molecules, because $Ca(NO_3)_2$ is *ionic*. One mole contains 1.807×10^{24} ions.

Question 38

What is the GEW of a metal if 46.4 g of it combines with 16.0 g of oxygen?

PART 5–GRAM ATOMIC WEIGHTS, MOLES AND FORMULAS   127

Answer 38

23.2 g

$$\text{OEIOU: } (GEW)_M = \frac{\# g_M}{\# g_O} \times (GEW)_O$$

Question 39

A 6.41-g sample of copper oxide is reduced with hydrogen to remove all oxygen from the oxide. If the copper residue weighs 5.10 g, what is the GEW of copper based on these data?

Answer 39

31.1 g

$$\text{OEIOU: } (GEW)_{Cu} = \frac{\# g\ Cu}{\# g\ CuO - \# g\ Cu} \times (GEW)_O$$

Question 40

Based on the data in Answer 39, what is the combining capacity of copper?

Answer 40

$n = 2$

$$\text{OEIOU: } n = \frac{GAW}{GEW}$$

(n is an integer; round off).

Question 41

The specific heat of a certain metal is found to be 0.0919 cal/g °C. What is its:

a. approximate GAW?

b. combining capacity, if its exact GEW is 31.77 g?

c. exact GAW?

Answer 41

a. $(GAW)_M \approx 65.3$ g

OEIOU: $(GAW)_M \approx \dfrac{6 \text{ cal/}°C}{\text{sp. heat}}$

b. $n = 2$

OEIOU: $n = \dfrac{(GAW)_M}{(GEW)_M}$

c. $(GAW)_M = 63.54$ g

OEIOU: $(GAW)_M = (GEW)_M \times n$

Question 42

Acetone has the empirical formula C_3H_6O. Its molecular weight is determined to be about 60. What is its molecular formula?

Answer 42

C_3H_6O

OEIOU: $MF = \dfrac{MW}{FW} \times EF$

Question 43

Ethyl acetate has the empirical formula C_2H_4O. If its molecular weight is found to be about 90, what is its molecular formula?

Answer 43

$C_4H_8O_2$

OEIOU: $MF = \dfrac{MW}{FW} \times EF$

Question 44

What is the empirical formula of N_2O_4?

PART 5–GRAM ATOMIC WEIGHTS, MOLES AND FORMULAS   129

Answer 44

NO_2

OEIOU: EF = MF/n

Question 45

What is the percentage composition of acetone, C_3H_6O, to three significant figures?

Answer 45

62.0% C, 10.4% H, 27.5% O

OEIOU: %C = $\dfrac{\text{awu C}}{\text{MW}} \times 100$

OEIOU: %H = $\dfrac{\text{awu H}}{\text{MW}} \times 100$

OEIOU: %O = $\dfrac{\text{awu O}}{\text{MW}} \times 100$

Question 46

A 3.00-g sample of an organic compound is found to contain 1.86 g of carbon and 0.312 g of hydrogen. What is its percentage composition?

Answer 46

62.0% C, 10.4% H, 27.6% O

OEIOU: %C = $\dfrac{\text{\# g C}}{\text{g sample}} \times 100$

OEIOU: %H = $\dfrac{\text{\# g H}}{\text{g sample}} \times 100$

OEIOU: %O = 100.00 ... % − (%C + %H)

Question 47

What is the empirical formula of a compound which is found to be 81.4% F, 17.2% C, and 1.44% H?

Answer 47

CHF$_3$

Step (1)

$$\text{OEIOU:} \quad \#\text{GAW}_C = \frac{\%\,C}{\text{GAW}_C}$$

(similar process for H and F)

Step (2)

$$\text{OEIOU:} \quad (C/H)\text{ ratio} = \frac{\#\text{GAW}_C}{\#\text{GAW}_H}$$

(similar process for F/H and H/H ratios)

Question 48

A 6.00-g sample of a yellow organic compound contains 4.38 g of fluorine, 1.03 g of carbon, and 0.0864 g of hydrogen. What is the empirical formula of the compound?

Answer 48

CHF$_3$

Step (1)

$$\text{OEIOU:} \quad \#\text{GAW}_C = \frac{\#\,g\,C}{\text{GAW}_C}$$

(similar process for H and F)

Step (2)

$$\text{OEIOU:} \quad (C/H) = \frac{\#\text{GAW}_C}{\#\text{GAW}_H}$$

(similar process for F/H and H/H ratios)

Question 49

A certain compound contains 24.8% C, 2.08% H, and 73.1% Cl. If its molecular weight is 98 ± 2, what is its molecular formula?

PART 5–GRAM ATOMIC WEIGHTS, MOLES AND FORMULAS

Answer 49

$C_2H_2Cl_2$

Find the empirical formula as in Answer 47. Then:

$$MF = \frac{MW}{FW} \times EF$$

PART 6
CHEMICAL EQUATIONS

Statement 1

A **chemical equation** is a **symbolic sentence** used to communicate what happens in a chemical reaction.

Example

$$CH_4 + 2O_2 \rightarrow CO_2 + 2H_2O$$

This statement might be "translated" to read: "One molecule of methane reacts with two molecules of oxygen to give one molecule of carbon dioxide and two molecules of water."

The **Law of Conservation of Matter** states that matter can be neither created nor destroyed in a chemical reaction, so that **coefficients** are used in front of formulas where necessary to assure that the same number of atoms of each kind appear on the left and right sides of the arrow.

Example of an Unbalanced Equation

$$CH_4 + O_2 \rightarrow CO_2 + H_2O$$

This statement contains on the left 1 C, 4 H's, and 2 O's; but on the right, 1 C, 2 H's, and 3 O's; therefore, it cannot be "true"; that is, it violates the Law of Conservation of Matter.

To **balance** this equation, place a 2 in front of O_2 on the left and a 2 in front of H_2O on the right:

$$CH_4 + 2O_2 \rightarrow CO_2 + 2H_2O$$

Now we have:

left side	right side
1 C	1 C
4 H	4 H
4 O	4 O

balanced

Question 1

Balance the equation:

$$Na + O_2 \rightarrow Na_2O$$

Answer 1

$$4Na + O_2 \rightarrow 2Na_2O$$

left	right
4 Na	4 Na
2 O	2 O

Question 2

Balance the equation:

$$NO_2 + H_2O \rightarrow HNO_3 + NO$$

Answer 2

$$3NO_2 + H_2O \rightarrow 2HNO_3 + NO$$

Question 3

Balance the equation:

$$C_2H_6 + O_2 \rightarrow CO_2 + H_2O$$

Answer 3

$$2C_2H_6 + 7O_2 \rightarrow 4CO_2 + 6H_2O$$

Statement 2

A chemical equation reveals not only the molecular (or atomic or ionic) ratios in which reactants react and products are produced, but also the precise **molar relationships** among the participants.

Example

$$CH_4 + 2O_2 \rightarrow CO_2 + 2H_2O$$

This statement may be translated to read: "One **mole** of methane reacts with two **moles** of oxygen to give one **mole** of carbon dioxide and two **moles** of water."

A **stoichiometric statement** (SS) is a mathematical equation relating the number of moles of one reactant or product to the number of moles of another reactant or product. It can be determined simply by examining the balanced equation.

PART 6—CHEMICAL EQUATIONS

For a generalized equation, $aA + bB \rightarrow cC + dD$, a typical SS would have the form:

$$\# \text{moles } A = \frac{a}{d} \times \# \text{moles } D$$

Other stoichiometric statements could be written relating the other substances involved in the reaction.

Question 4

Write the stoichiometric statement relating CH_4 and O_2 in the equation $CH_4 + 2O_2 \rightarrow CO_2 + 2H_2O$

Answer 4

$$\# \text{moles } CH_4 = \frac{1}{2} \times \# \text{moles } O_2$$

Defining equation: $\# \text{moles } A = \frac{a}{c} \times \# \text{moles } C$

SS: $\# \text{moles } CH_4 = \frac{1}{2} \times \# \text{moles } O_2$

Note: The correct SS can always be obtained by using the relationship given in Statement 2. It is more important, however, to *think* about why this turns out to be correct. We may "translate" the SS as follows: "The number of moles of methane is *half* the number of moles of oxygen." From the coefficients in the chemical equation, we can see that this is a true statement.

Question 5

Write the stoichiometric statement relating O_2 and H_2O for the equation $CH_4 + 2O_2 \rightarrow CO_2 + 2H_2O$

Answer 5

$\# \text{moles } O_2 = \# \text{moles } H_2O$

$$\# \text{moles } O_2 = \frac{2}{2} \times \# \text{moles } H_2O$$

$$= \# \text{moles } H_2O$$

Statement 3

Knowing that # moles A = #g A/GMW_A, it is easy to determine **weight relationships** of participants in a reaction.

Knowing the quantity of any one reactant or product in a reaction, we can start with the stoichiometric statement and calculate the corresponding quantity of any other reactant or product.

The expected (theoretical) quantity of product calculated in this way from a given quantity of reactant is called the **theoretical yield** of that product.

Question 6

"Burning" is a chemical reaction between oxygen and another reactant. When 32.0 g of methane, CH_4, is burned to give CO_2 and H_2O, what is the theoretical yield (in grams) of CO_2?

Answer 6

88.0 g

Balanced equation: $CH_4 + 2O_2 \rightarrow CO_2 + 2H_2O$

SS: # moles CO_2 = # moles CH_4

Substituting: $\dfrac{\text{\#g } CO_2}{GMW_{CO_2}} = \dfrac{\text{\#g } CH_4}{GMW_{CH_4}}$

OEIOU: #g CO_2 = $\dfrac{\text{\#g } CH_4}{GMW_{CH_4}} \times GMW_{CO_2}$

$= \dfrac{32.0 \text{ g}}{16.0 \text{ g/mole}} \times 44.0 \text{ g/mole}$

Question 7

What is the theoretical yield of H_2O when 32.0 g of methane is burned?

Answer 7

72.0 g

SS: # moles H_2O = 2 × # moles CH_4

OEIOU: #g H_2O = 2 × $\dfrac{\text{\#g } CH_4}{GMW_{CH_4}} \times GMW_{H_2O}$

Question 8

How many grams of O_2 would be used up in burning 32.0 g of methane?

Answer 8

128 g

SS: # moles O_2 = 2 × # moles CH_4

OEIOU: #g O_2 = 2 × $\dfrac{\text{\#g } CH_4}{GMW_{CH_4}}$ × GMW_{O_2}

Question 9

When heated, barium peroxide (BaO_2) decomposes to give barium oxide (BaO) and oxygen. Given 16.9 g of BaO_2:

a. how much BaO could one produce?

b. what would be the theoretical yield of O_2?

Answer 9

a. 15.3 g BaO

b. 1.60 g O_2

Question 10

An 8.00-g sample of a compound containing only C, H, and O is burned to give 19.20 g of CO_2 and 3.92 g of water. What is its empirical formula?

Answer 10

C_3H_3O

To calculate percentage composition:

$$\%\,C = \dfrac{\text{\#g C}}{\text{g sample}} \times 100$$

$$= \dfrac{\text{\#g } CO_2 \times \dfrac{GAW_C}{GMW_{CO_2}} \times 100}{\text{g sample}}$$

$$= 65.6\%$$

PART 6—CHEMICAL EQUATIONS   137

$$\% H = \frac{\# g\ H}{g\ sample} \times 100$$

$$= \frac{2 \times \# g\ H_2O \times \frac{GAW_H}{GMW_{H_2O}}}{g\ sample} \times 100$$

$$= 5.44\%\ H$$

$$\% O = 100.00\% - (\% C + \% H)$$

$$= 29.0\ \% O$$

Then, as in Part 5, Answer 25, find the empirical formula, C_3H_3O.

Statement 4

In practice, pure product from a reaction is never as great as the theoretical yield predicts. That is, the **actual yield** is always less than 100% of the theoretical yield. To calculate the actual **percentage yield**, use the relationship:

$$\% \text{ yield} = \frac{\text{actual yield}}{\text{theoretical yield}} \times 100$$

where actual and theoretical yield may both be expressed in any convenient units (moles, grams, pounds, and so forth).

Question 11

The actual yield of oxygen from the heating of 16.93 g of BaO_2 is 0.800 g. What is the percentage yield of O_2? (See Answer 9b. for the theoretical yield of O_2.)

Answer 11

50.0%

$$\% \text{ yield} = \frac{g_{Act}}{g_{Theor}} \times 100$$

$$= \frac{0.800\ g}{1.60\ g} \times 100$$

$$= 50.0\%$$

Question 12

The actual yield of CO_2 obtained on burning 48.0 g of methane was 100.0 g.

a. Show OEIOU for calculating percentage yield of CO_2.
b. Calculate the percentage yield of CO_2.

Answer 12

a. $\%\text{ yield}_{CO_2} = \dfrac{\#g_{Act}}{\#g_{CH_4}} \times \dfrac{GMW_{CH_4}}{GMW_{CO_2}} \times 100$

Defining equation: $\%\text{ yield} = \dfrac{\#g_{Act}}{\#g_{Theor}} \times 100$

Substituting: $\%\text{ yield} = \dfrac{\#g_{Act}}{\#g\ CH_4 \times \dfrac{GMW_{CO_2}}{GMW_{CH_4}}} \times 100$

$= \dfrac{\#g_{Act}}{\#g_{CH_4}} \times \dfrac{GMW_{CH_4}}{GMW_{CO_2}} \times 100$

b. 75.8%

Statement 5

Chemical equations specify exactly how much of one reactant is required to "take care of" a given amount of any other reactant. When one reactant is present in **excess**, the excess will not react because there is not enough of the other reactant to "take care of it."

The reactant present in the **smallest quantity relative to that specified by the chemical equation** is called the **limiting reagent** inasmuch as it limits the theoretical yield.

Question 13

Three moles of CH_4 are burned in five moles of O_2. Which compound is the limiting reagent?

Answer 13

oxygen

From the equation

$$CH_4 + 2O_2 \rightarrow CO_2 + 2H_2O$$

we see that one mole of CH_4 requires 2 moles of O_2.

SS: # moles O_2 = 2 × moles CH_4

Thus 3 moles of CH_4 requires 6 moles of O_2. Since only 5 moles are available, O_2 *limits* the theoretical yield of CO_2 and H_2O.

Question 14

If 16.01 g of methane is burned in 96.00 g of oxygen, which compound is the limiting reagent?

Answer 14

methane

From the equation

$$CH_4 + 2O_2 \rightarrow CO_2 + 2H_2O$$

we know that 16.01 g (1 mole) of CH_4 requires 64.00 g (2 moles) of O_2. We have 96.00 g of O_2, which is *3 moles*, but only 1 mole of CH_4. Thus the CH_4 limits the theoretical yield in this case.

Question 15

How much unreacted oxygen is left over in the burning of methane described in Question 13?

Answer 15

1.000 mole (or 32.00 g) of O_2

The 1 mole of CH_4 reacted with exactly 2 moles of O_2, so we are left with an excess of 3 moles − 2 moles = 1 mole of O_2.

Question 16

What is the limiting reagent when 8.01 g of methane is burned in 30.0 g of oxygen?

Answer 16

oxygen

TESTING YOUR MASTERY—Part 6 Chemical Equations

Question 17

Balance the equation:

$$P_4 + Cl_2 \rightarrow PCl_3$$

Answer 17

$$P_4 + 6Cl_2 \rightarrow 4PCl_3$$

Question 18

Balance the equation:

$$H_2 + Fe_3O_4 \rightarrow Fe + H_2O$$

Answer 18

$$4H_2 + Fe_3O_4 \rightarrow 3Fe + 4H_2O$$

Question 19

Balance the equation:

$$KClO_3 \rightarrow KCl + O_2$$

Answer 19

$$2KClO_3 \rightarrow 2KCl + 3O_2$$

Question 20

$$2KClO_3 \rightarrow 2KCl + 3O_2$$

Write the stoichiometric statements relating the following substances in the above balanced equation:

a. $KClO_3$ and KCl

b. $KClO_3$ and O_2

c. KCl and O_2

Answer 20

a. # moles $KClO_3$ = # moles KCl

b. # moles $KClO_3$ = $\frac{2}{3}$ × # moles O_2

c. # moles KCl = $\frac{2}{3}$ × # moles O_2

Question 21

Write the stroichiometric statement relating the oxygen used to the CO_2 produced when ethane, C_2H_6, is burned.

Answer 21

moles O_2 = $\frac{7}{4}$ × # moles CO_2

Balanced equation: $2C_2H_6 + 7O_2 \rightarrow 4CO_2 + 6H_2O$

Question 22

What is the theoretical yield of water (a) in moles, and (b) in grams, when 45.0 g of ethane is burned?

Answer 22

a. 4.49 moles

OEIOU: # moles H_2O = 3 × $\dfrac{\text{\# g } C_2H_6}{GMW_{C_2H_6}}$

b. 80.9 g

OEIOU: # g H_2O = 3 × $\dfrac{\text{\# g } C_2H_6}{GMW_{C_2H_6}}$ × GMW_{H_2O}

Question 23

How much water (in grams) is required to react with 9.66 g of Na metal according to the following equation:

$2Na + 2H_2O \rightarrow H_2 + 2NaOH$

PART 6–CHEMICAL EQUATIONS

Answer 23

7.56 g

OEIOU: $\#g\ H_2O = \dfrac{\#g\ Na}{GAW_{Na}} \times GMW_{H_2O}$

Question 24

Hydrogen sulfide reacts with zinc metal to give zinc sulfide and H_2. Beginning with 16.3 g of zinc,

a. how much zinc sulfide could be produced?

b. what is the percentage yield of zinc sulfide, if 19.7 g of the sulfide is actually collected?

Answer 24

a. 24.3 g

OEIOU: $\#g\ ZnS = \dfrac{\#g\ Zn}{GAW_{Zn}} \times GMW_{ZnS}$

b. 81.1%

OEIOU: Percentage yield $= \dfrac{act\ yield}{theor\ yield} \times 100$

(where "theor yield" = $\#g\ ZnS$ from a. above)

Question 25

The actual yield of CO_2 collected when 20.0 g of ethane, C_2H_6, was burned was 50.3 g. What was the percentage yield of CO_2 in this case?

Answer 25

85.9%

OEIOU: Percentage yield $= \dfrac{(\#g\ CO_2)_{act}}{2 \times \#g\ C_2H_6} \times \dfrac{GMW_{C_2H_6}}{GMW_{CO_2}} \times 100$

Question 26

Sodium reacts with chlorine to give sodium chloride. If 50.0 g of sodium metal is mixed with 70.9 g of chlorine,

a. which is the limiting reagent?
b. how many grams of the excess reagent will remain unreacted?

Answer 26

a. Cl_2 is the limiting reagent, since 50.0 g of Na requires 77.1 g of Cl_2 to react completely, and only 70.9 g is available.

b. 4.0 g Na excess.

OEIOU: $(\#g\,Na)_{exc} = \left[\dfrac{2 \times \#g\,Cl_2}{GMW_{Cl_2}} \times GAW_{Na}\right] - (\#g\,Na)_{act}$

Question 27

A 7.00-g sample of an organic liquid containing only C, H, and O is burned to give 14.0 g of CO_2 and 5.70 g of H_2O. What is the empirical formula of the compound?

Answer 27

C_2H_4O

Step 1

OEIOU: $\%\,C = \dfrac{\#g\,CO_2}{\#g\,sample} \times \dfrac{GAW_C}{GMW_{CO_2}} \times 100$

OEIOU: $\%\,H = 2 \times \dfrac{\#g\,H_2O}{\#g\,sample} \times \dfrac{GAW_H}{GMW_{H_2O}} \times 100$

OEIOU: $\%\,O = 100.0\ldots\% - (\%C + \%H)$

Step 2

OEIOU: $\#\,GAW_C = \dfrac{\%C}{GAW_C}$

(similar procedure for H and O)

Step 3

OEIOU: ratio of (C/O) = $\dfrac{\#\,GAW_C}{\#\,GAW_O}$

(similar procedure for H/O and O/O ratios)

PART 7
HEAT FLOW DURING CHEMICAL REACTIONS

Statement 1

We know from experience that during some chemical reactions the reaction mixture warms up, while in others it cools off. This is evidence that **heat flow** has taken place. Such changes can be studied quantitatively by measuring the amount of heat, Q, gained or lost by the system.

When heat flows **out** of a system into the surroundings, the process is said to be **exothermic** $(-Q)$.

When heat flows **into** a system from the surroundings, the process is said to be **endothermic** $(+Q)$.

Question 1

When a mole of Mg metal is burned in a bomb calorimeter (a closed system used in studies of this type) at 25°C and 1 atm of pressure, 143.8 kcal of heat must be removed to return the MgO produced to 25°C. Is the reaction endothermic or exothermic?

> **Answer 1**
>
> Exothermic.
>
> Heat was *removed* to restore the original temperature. This means that heat was *given off* during the reaction, warming up the MgO produced.

Question 2

What is the value of Q for the burning of a mole of Mg metal at 25°C and 1 atm of pressure? (see Question 1.)

> **Answer 2**
>
> $Q = -143.8$ kcal/mole
>
> > The minus sign indicates that the heat must be *removed* to restore the original temperature.

Question 3

When a mole of N_2 is reacted with oxygen at 25°C and 1 atm to form NO gas, the surroundings absorb 21.6 kcal of heat.

a. Is the reaction exothermic or endothermic?

b. What is the value of Q when one mole of N_2 is burned to give NO?

Answer 3

a. endothermic

   Heat flowed from the surroundings *into* the system.

b. +21.6 kcal/mole

   The + sign indicates that the reaction is *endothermic*.

Statement 2

The experimental determination of Q for a chemical reaction is accomplished by carrying out the reaction in an insulated **calorimeter** and finding the heat gained by or lost to the calorimeter from the reaction vessel contained in it. This heat gain or loss may be found by multiplying the **heat capacity**, C, of the calorimeter by the change in its temperature as the reaction takes place. (The heat capacity is the amount of heat required to raise the temperature of the calorimeter by one degree Celsius.) Mathematically:

$$Q = C \Delta t = C(t_2 - t_1)$$

The heat capacity of the calorimeter is obtained in a separate experiment by carrying out a reaction whose Q is known, measuring the temperature change, Δt, and calculating C; or by supplying a known Q from a heater, measuring t, and calculating C.

Question 4

The heat of combustion of benzoic acid is -6.315 kcal/g. When 2.00 g of benzoic acid is burned in a certain calorimeter, the temperature of the calorimeter rises by 5.00°C. What is C for the calorimeter?

Answer 4

-2.53 kcal/°C

Defining equation: $Q = C \Delta t$ or
$C = Q/\Delta t$

OEIOU: $C = \dfrac{\text{(heat of comb.) (\# g)}}{\Delta t}$

$= \dfrac{(-6.315 \text{ kcal/g}) (2.00 \text{ g})}{5.00°C}$

$= 2.53 \text{ kcal/}°C$

Question 5

When 3.00 g of an organic compound is burned in the calorimeter used in Question 3 above, the temperature of the calorimeter rises by 4.00°C. What is the heat of combustion (burning) of the compound in kcal/g?

Answer 5

-3.37 kcal/g

OEIOU: heat of comb. $= \dfrac{C\Delta t}{\#g}$

Statement 3

A typical calorimeter consists of a reaction vessel submerged in water held in an insulated container. When the reaction takes place, heat flows either into the water from the reaction vessel, or from the water into the reaction vessel.

By measuring the temperature change, Δt, for this known weight of water, g_{H_2O}, we know how much heat has been gained or lost by the water:

$$Q_{H_2O} = c_{H_2O} \times \Delta t \times g_{H_2O}$$

The specific heat of water, c_{H_2O}, is 1.000 cal/g°C.

The Law of the Conservation of Energy states that the heat gained (or lost) by the water and the calorimeter apparatus (container and vessel) must equal the heat lost (or gained) by the reaction system:

$$(Q_{H_2O} + Q_{app}) = -Q_{rx}$$

For convenience, the heat capacity of the apparatus is generally expressed in terms of its "**water equivalent**," the weight of water equivalent to the apparatus in heat capacity. Thus:

$$c_{H_2O} \times (t_2 - t_1) \times (g_{H_2O} + g_{WE}) = -Q_{rx}$$

Question 6

When 0.121 g of Mg metal is burned in excess oxygen inside a bomb calorimeter, the 300.0 g of water increases in temperature from 25.00°C to 26.44°C. The water equivalent for the apparatus is 201 g.

a. What is Q for the process?

b. What is the heat of combustion in kcal/mole for Mg metal?

Answer 6

a. $Q_{rx} = -721$ cal

$Q_{rx} = -(Q_{H_2O} + Q_{app})$

$= c_{H_2O} \times (t_2 - t_1) \times (g_{H_2O} + g_{WE})$

b. $Q_{comb} = -145$ kcal/mole

$\#a = \#b \times$ C.F.

$\# \dfrac{\text{kcal}}{\text{mole}} = \#(\text{cal/g}) \times \dfrac{\text{kcal/mole}}{\text{cal/g}}$

$= \dfrac{\#\,\text{cal}}{\#\,\text{g}} \times \dfrac{\text{kcal}}{\text{cal}} \times \dfrac{\text{g}}{\text{mole}}$

$= \dfrac{-721\,\text{cal}}{0.121\,\text{g}} \times \dfrac{1\,\text{kcal}}{1000\,\text{cal}} \times \dfrac{24.31\,\text{g}}{1\,\text{mole}}$

$= -145$ kcal/mole

Statement 4

The heat flow in a chemical reaction is **directly proportional** to the number of moles of substance involved in the reaction. Mathematically:

$Q = k \times \#\text{moles substance}$

where k is a proportionality constant characteristic of a given reaction.

Question 7

The reaction of gaseous H_2 with one mole of gaseous N_2 to give ammonia, NH_3, is exothermic by 22.0 kcal. What is the value of Q when 56.0 g of N_2 reacts completely to produce ammonia?

Answer 7

$Q = -44.0$ kcal

Balanced equation: $3H_2 + N_2 \rightarrow 2NH_3$

$$Q = -22.0 \text{ kcal}$$

Defining equation: $Q = k \times \# \text{moles } N_2$

$$= k \times \frac{\# g\, N_2}{GMW_{N_2}}$$

From data for *one mole* of N_2, we know:

$$k = (Q/\# \text{moles } N_2)_{1 \text{ mole}}$$

Thus:

OEIOU: $Q = (Q/\# \text{moles } N_2)_{1 \text{ mole}} \times \dfrac{\# g\, N_2}{GMW_{N_2}}$

$$= \frac{-22.0 \text{ kcal}}{1 \text{ mole}} \times \frac{56.0 \text{ g}}{28.0 \text{ g/mole}}$$

$$= -44.0 \text{ kcal}$$

Question 8

What is the value of Q when 9.06 g of H_2 reacts completely with N_2 to give ammonia? (See Question 6.)

Answer 8

-33.0 kcal

Defining equation:

OEIOU: $Q = \left(\dfrac{Q}{\# \text{moles } H_2}\right)_{3 \text{ moles}} \times \dfrac{\# g\, H_2}{GMW_{H_2}}$

$$= \frac{-22.0 \text{ kcal}}{3.00 \text{ moles}} \times \frac{9.06 \text{ g}}{2.01 \text{ g/mole}}$$

$$= -33.0 \text{ kcal}$$

Question 9

What is the value of Q when 51.0 g of ammonia is produced from the reaction of H_2 with N_2? (See Question 6.)

PART 7—HEAT FLOW DURING CHEMICAL REACTIONS

Answer 9

−33.0 kcal

Defining equation:

$$\text{OEIOU:}\quad Q = \frac{Q}{\#\text{ moles NH}_3} \times \frac{\#\text{g NH}_3}{\text{GMW}_{\text{NH}_3}}$$

Statement 5

When the **direction** of a reaction is **reversed**, the **sign of Q** is reversed.

Question 10

What is the value of Q for the process

$$2\text{NH}_3 \rightarrow 3\text{H}_2 + \text{N}_2$$

(See Question 6.)

Answer 10

$Q = +22.0$ kcal

We know from Question 6 that in order for N_2 reacting with H_2 to produce NH_3, Q must equal −22.0 kcal, so for the *reverse* process Q must be +22.0 kcal.

Question 11

For the process $2\text{MgCl}_2 + \text{O}_2 \rightarrow 2\text{MgO} + \text{Cl}_2$, Q is +19.2 kcal. What is Q for the reaction of 70.9 g of chlorine with MgO to give MgCl_2 and oxygen?

Answer 11

−9.60 kcal

The reaction of Cl_2 with MgO is the *reverse* of the process for which Q is given; hence for 2 moles of Cl_2 reacting with 2 moles of MgO, $Q = -19.2$ kcal. Therefore:

$$Q = k \times \#\text{ moles Cl}_2$$

$$= \left(\frac{Q}{\#\text{ moles Cl}_2}\right)_{2\text{ moles}} \times \frac{\#\text{g Cl}_2}{\text{GMW}_{\text{Cl}_2}}$$

$$= \frac{-19.2\text{ kcal}}{2.00\text{ moles}} \times \frac{70.9\text{ g}}{70.9\text{ g/mole}}$$

$$= -9.60\text{ kcal}$$

Statement 6

Hess's Law states that when a reaction can be written as the **sum** of two or more reactions, $Q_{overall}$ will be given by adding Q for each reaction included in the summation. Mathematically:

$$Q_{ov} = Q_A + Q_B + \cdots$$

Question 12

Given:

$2HCl \rightarrow H_2 + Cl_2$

$Q_1 = +44.2$ kcal

$2H_2 + O_2 \rightarrow 2H_2O$

$Q_2 = -136.6$ kcal

What is Q for the reaction:

$$4HCl + O_2 \rightarrow 2Cl_2 + 2H_2O$$

Answer 12

-48.2 kcal

Summation to give the reaction of interest:

$2(2HCl \rightarrow H_2 + Cl_2)$
$2H_2 + O_2 \rightarrow 2H_2O$
_____
$4HCl + O_2 \rightarrow 2Cl_2 + 2H_2O$

Defining equation: $Q_{ov} = Q_A + Q_B$

OEIOU: $Q_{ov} = 2Q_1 + Q_2$

$= 2(44.2 \text{ kcal}) + (-136.6 \text{ kcal})$

$= -48.2$ kcal

Question 13

Given:

$S + O_2 \rightarrow SO_2 \qquad Q_1 = -71.0$ kcal

$2S + 3O_2 \rightarrow 2SO_3 \qquad Q_2 = -189.0$ kcal

What is Q for the reaction:

$$2SO_2 + O_2 \rightarrow 2SO_3$$

Answer 13

−47.0 kcal

Summation:

$$2(SO_2 \rightarrow O_2 + S)$$
$$2S + 3O_2 \rightarrow 2SO_3$$
$$\overline{2SO_2 + O_2 \rightarrow 2SO_3}$$

Defining equation: $Q_{ov} = Q_A + Q_B$

OEIOU: $Q_{ov} = -2Q_1 + Q_2$

$\qquad = -2(-71.0 \text{ kcal}) + (-189.0 \text{ kcal})$

$\qquad = -47.0 \text{ kcal}$

Question 14

Given:

$2H_2 + O_2 \rightarrow 2H_2O \qquad Q_1 = -136.6 \text{ kcal}$

$N_2 + 3H_2 \rightarrow 2NH_3 \qquad Q_2 = -22.0 \text{ kcal}$

What is Q for the process:

$$4NH_3 + 3O_2 \rightarrow 2N_2 + 6H_2O$$

Answer 14

−365.8 kcal

Summation:

$$3(2H_2 + O_2 \rightarrow 2H_2O)$$
$$2(2NH_3 \rightarrow N_2 + 3H_2)$$
$$\overline{4NH_3 + 3O_2 \rightarrow 6H_2O + 2N_2}$$

OEIOU: $Q_{ov} = 3Q_1 + (-2Q_2)$

Statement 7

The standard molar **heat of formation** of a substance, ΔH_f°, is the heat flow associated with the formation of one mole of a compound from its elements in their **standard states**—usually 25°C and 1 atm pressure. The standard heat of formation of any element is taken to be **zero**.

PART 7—HEAT FLOW DURING CHEMICAL REACTIONS

When the heats of formation of compounds involved in a reaction are known, the **heat of reaction** may be calculated by subtracting the sum of the heats of formation of the reactants from the sum of the heats of formation of the products. Mathematically:

$$Q = \Delta H_f^\circ, \text{products} - \Delta H_f^\circ, \text{reactants}$$

Question 15

For $H_2O(l)$ ΔH_f° is -68.3 kcal/mole; for $SO_3(g)$ it is -94.5 kcal/mole; for $H_2SO_4(l)$ it is -193.9 kcal/mole. What is the heat of reaction for the process:

$$H_2O(l) + SO_3(g) \rightarrow H_2SO_4(l)$$

Answer 15

-31.1 kcal

Defining equation: $\Delta H_{rx} = \Delta H_{f,pr}^\circ - \Delta H_{f,re}^\circ$

OEIOU: $\Delta H_{rx} = \Delta H_{f,H_2SO_4}^\circ - (\Delta H_{f,H_2O}^\circ + \Delta H_{f,SO_3}^\circ)$

Question 16

Two moles of SO_2 gas are burned to give SO_3 gas. The heat of formation for $SO_2(g)$ is -71.0 kcal/mole; for $SO_3(g)$ it is -94.5 kcal/mole. What is the heat of reaction for the burning process:

$$2SO_2(g) + \tfrac{1}{2}O_2(g) \rightarrow 2SO_3(g)$$

Answer 16

-47.0 kcal

OEIOU: $\Delta H_{rx} = 2\Delta H_{f,SO_3}^\circ - (2\Delta H_{f,SO_2}^\circ + \tfrac{1}{2}\Delta H_{f,O_2}^\circ)$

$= 2(-94.5 \text{ kcal}) - 2(-71.0 \text{ kcal}) - \tfrac{1}{2}(0 \text{ kcal})$

$= -47.0 \text{ kcal}$

Question 17

The heat of reaction for the process shown below is $+23.5$

kcal/mole of H_2O. What is ΔH_f° for hydrogen peroxide, $H_2O2(l)$?

$$H_2O(l) + \tfrac{1}{2}O_2(g) \rightarrow H_2O_2(l)$$

Answer 17

−44.8 kcal

OEIOU: $\Delta H_{f,H_2O_2}^\circ = \Delta H_{rx} + (\Delta H_{f,H_2O}^\circ + \tfrac{1}{2}\Delta H_{f,O_2}^\circ)$

Question 18

For the reduction of Al_2O_3 by H_2 gas (see below) the heat of reaction is +194.2 kcal/mole of Al_2O_3. If the heat of formation of liquid H_2O is −68.3 kcal/mole, what is $\Delta H_{f,Al_2O_3(s)}^\circ$?

$$Al_2O_3(s) + 3H_2(g) \rightarrow 2Al(s) + 3H_2O(l)$$

Answer 18

−399.1 kcal

OEIOU: $\Delta H_{rx} = (2\Delta H_{f,Al}^\circ + 3\Delta H_{f,H_2O}^\circ)$
$- (\Delta H_{f,Al_2O_3}^\circ + \Delta H_{f,H_2}^\circ)$

TESTING YOUR MASTERY—Part 7
Heat Flow During Chemical Reactions

Question 19

When 3.00 g of sucrose is burned in a certain calorimeter, the temperature of the calorimeter rises from 25.11°C to 29.37°C. What is the heat capacity of the calorimeter if the heat of combustion of sucrose is −4.02 kcal/g?

Answer 19

−2.83 kcal/°C

OEIOU: $C = \dfrac{Q_{comb} \times \#g}{t_2 - t_1}$

Question 20

A 2.58-g sample of cyclohexane, C_6H_{12} is burned in a calorimeter with a heat capacity of -2.08 kcal/°C. If the temperature of the calorimeter rises by 13.9°C, what is the heat of combustion of cyclohexane in:

a. kcal/g ?

b. kcal/mole ?

Answer 20

a. -11.2 kcal/g.

OEIOU: $Q_{comb} = \dfrac{C \Delta t}{\#g}$

b. -941 kcal/mole

OEIOU: $Q_{comb} = \dfrac{C \Delta t}{\#g} \times GMW_{C_6H_{12}}$

Question 21

A certain bomb calorimeter contains 520.0 g of water. When 0.400 of graphite (carbon) is burned in excess oxygen in the calorimeter, the water temperature increases by 3.04°C. The water equivalent of the apparatus is 164 g.

a. What is the heat flow, Q, for the burning process?

b. What is the heat of combustion of graphite in kcal/mole based on these data?

Answer 21

a. -3.12 kcal

OEIOU: $Q_{rx} = C_{H_2O} \times \Delta t \times (g_{H_2O} + g_{WE})$

b. -93.7 kcal/mole

OEIOU: $Q_{comb} = Q_{rx} \times \dfrac{GAW_c}{\#g_c}$

Question 22

The reaction of CH_4 and Cl_2 to give CH_3Cl and HCl is exothermic by 25.6 kcal/mole CH_4. What is the value of Q when

PART 7—HEAT FLOW DURING CHEMICAL REACTIONS

12.0 g of CH_4 reacts completely to give CH_3Cl, methyl chloride?

Answer 22

$Q_2 = -19.2$ kcal

OEIOU: $Q_2 = \dfrac{Q_1}{(\text{\# moles } CH_4)_1} \times \dfrac{\text{\# g}_2}{GMW_{CH_4}}$

Question 23

When 4.86 g of CH_4 reacts with Cl_2 to give carbon tetrachloride, CCl_4, the reaction releases 29.5 kcal of heat.

a. What would Q be when 16.0 g of CH_4 is converted into CCl_4?

b. What is Q_{rx} for this process in kcal/mole CH_4?

Answer 23

a. $Q_2 = -97.1$ kcal

OEIOU: $Q_2 = Q_1 \times \dfrac{\text{\# g}_2}{\text{\# g}_1}$

b. $Q_{rx} = -97.1$ kcal/mole

OEIOU: $Q_{rx} = \dfrac{Q_2}{\text{\# g}_2} \times GMW_{CH_4}$

Question 24

Gaseous NH_3 and HCl react to give solid NH_4Cl and to release 12.8 kcal of heat per mole of NH_3. What is Q for the process:

$$NH_4Cl \rightarrow NH_3 + HCl$$

Answer 24

$Q_2 = +12.8$ kcal

OEIOU: $Q_2 = -Q_1$

Question 25

If Q is -28.6 kcal per mole for $PCl_3 + Cl_2 \rightarrow PCl_5$, what is Q when 52.1 g of PCl_5 decomposes into PCl_3 and Cl_2?

Answer 25

$Q_2 = +7.16$ kcal

OEIOU: $Q_2 = \dfrac{-Q_1}{(\# \text{mole } PCl_5)_1} \times \dfrac{(\# \text{g } PCl_5)_2}{GMW_{PCl_5}}$

Question 26

Given:

$CH_4 + Cl_2 \rightarrow CH_3Cl + HCl \quad Q_1 = -25.6$ kcal

$CH_4 + 4Cl_2 \rightarrow CCl_4 + 4HCl \quad Q_2 = -97.1$ kcal

What is Q for the following reaction:

$2CH_3Cl + 6Cl_2 \rightarrow 2CCl_4 + 6HCl$

Answer 26

$Q_{ov} = -143$ kcal

OEIOU: $Q_{ov} = -2Q_1 + 2Q_2$

Question 27

If for $N_2 + O_2 \rightarrow 2NO$, Q is endothermic by 21.6 kcal; and if for $2NO + O_2 \rightarrow 2NO_2$, Q_2 is exothermic by 27.0 kcal; what is Q for the following:

$N_2 + 2O_2 \rightarrow 2NO_2$

Answer 27

-5.4 kcal

OEIOU: $Q_{ov} = Q_1 + Q_2$

Question 28

Given:

$$C_2H_4 + 3O_2 \rightarrow 2CO_2 + 2H_2O$$

$Q_1 = -202$ kcal

$$2H_2 + O_2 \rightarrow 2H_2O$$

$Q_2 = -137$ kcal

$$2C_2H_6 + 7O_2 \rightarrow 4CO_2 + 6H_2O$$

$Q_3 = -475$ kcal

What is Q for the following reaction:

$$C_2H_4 + H_2 \rightarrow C_2H_6$$

Answer 28

$Q_{ov} = -33$ kcal

OEIOU: $Q_{ov} = Q_1 + \frac{1}{2}Q_2 - \frac{1}{2}Q_3$

Question 29

Given the following heats of formation:

$\Delta H^\circ_{f,NO} = +21.6$ kcal/mole

$\Delta H^\circ_{f,NO_2} = +8.1$ kcal/mole

calculate the heat of reaction for:

$$2NO(g) + O_2(g) \rightarrow 2NO_2(g)$$

Answer 29

$\Delta H_{rx} = -27.0$ kcal

OEIOU: $\Delta H_{rx} = 2\Delta H^\circ_{f,NO_2} - 2\Delta H^\circ_{f,NO} - \Delta H^\circ_{f,O_2}$

Question 30

For NO(g), ΔH°_f is +21.6 kcal/mole; for $NO_2(g)$ ΔH°_f is +8.1 kcal/mole; for $SO_2(g)$, ΔH°_f is -71.0 kcal/mole; and for $SO_3(g)$, ΔH°_f is -94.4 kcal/mole. What is the heat of reaction for:

$$NO_2(g) + SO_2(g) \rightarrow NO(g) + SO_3(g)$$

Answer 30

$\Delta H_{rx} = -9.9$ kcal

OEIOU: $\Delta H_{rx} = \Delta H^°_{f,NO} + \Delta H^°_{f,SO_3} - \Delta H^°_{f,NO_2}$

$\quad\quad\quad\quad\quad - \Delta H^°_{f,SO_2}$

Question 31

The heat of reaction is -31.2 kcal for the reduction of CuO by H_2:

$$CuO(s) + H_2(g) \rightarrow Cu(s) + H_2O(l)$$

If the heat of formation for liquid H_2O is -68.3 kcal/mole, what is $\Delta H^°_f$ for CuO?

Answer 31

-37.1 kcal/mole

OEIOU: $\Delta H^°_{f,CuO} = \Delta H^°_{f,H_2O} + \Delta H^°_{f,Cu} - \Delta H^°_{f,H_2} - \Delta H^°_{rx}$

PART 8
THE FIRST LAW OF THERMODYNAMICS

Statement 1

According to the law of the conservation of energy, as a system changes state, energy may be converted from one form into another but may be neither created nor destroyed. This principle is also implied in the **First Law of Thermodynamics**, whose mathematical statement is as follows:

$$\Delta E = Q - W$$

ΔE is the symbol used for the change in **internal energy** of the system, the name given to the total energy of the system. Q is the **heat flow** taking place and W the **work** done when the change of state occurs. When the First Law is expressed by the above equation, we use these sign conventions:

$-Q$	heat given off	$+W$	work done by system
$+Q$	heat absorbed	$-W$	work done on system

Question 1

When one mole of liquid H_2O at 100°C and 1 atm of pressure is converted to gaseous H_2O at 100°C and 1 atm, 9.7 kcal of heat are absorbed by the system and 0.73 kcal of work done in expanding against the atmospheric pressure. What is ΔE for the process?

Answer 1

+9.0 kcal

OEIOU: $\Delta E = Q - W$
   $= +9.7$ kcal $- (+0.73$ kcal$)$
   $= +9.0$ kcal

Question 2

When one mole of water vapor at 100°C and 1 atm pressure is converted into liquid water at 100°C and 1 atm, 9.7 kcal of heat are given off by the system and 0.73 kcal of work done on the system by atmospheric pressure contracting it. What is ΔE for the process?

Answer 2

−9.0 kcal

OEIOU: $\Delta E = Q - W$

$= -9.7 \text{ kcal} - (-0.73 \text{ kcal})$

$= -9.0 \text{ kcal}$

Statement 2

The most usual kind of work done during changes in state is **pressure-volume** work; that is, work done by the system in pushing against confining (usually atmospheric) pressure when a system **expands**, or work done on the system by the surroundings in **contracting** the system so that it takes up a smaller volume in its final state. For the process at a constant pressure:

$$\text{system } (p_1, v_1) \rightarrow \text{system } (p_1, v_2)$$

$$W = p\Delta V = p(V_2 - V_1)$$

In a typical chemical reaction, the pressure is simply atmospheric pressure.

Question 3

A certain gas with a volume of 4.0 liters at 1 atm pressure, when cooled at constant pressure, contracts to occupy 2.5 liters. What is W for the change?

Answer 3

−1.5 L-atm

OEIOU: $W = p(V_2 - V_1)$

$= (1 \text{ atm})(2.5 \text{ L} - 4.0 \text{ L})$

$= -1.5 \text{ L-atm}$

Question 4

When a certain chemical reaction takes place at 1 atm of pressure, the products occupy 45.3 L, whereas the reactants occupied only 15.1 L. What is W for the reaction?

Answer 4

+30.2 L-atm

$W = p(V_2 - V_1)$

$= (1 \text{ atm})(45.3 \text{ L} - 15.1 \text{ L})$

$= +30.2 \text{ L-atm}$

Statement 3

One **liter-atmosphere** (L-atm) of work is the equivalent of **24.2 calories**.

Question 5

What is W in calories when:

a. $W = -1.5$ L-atm?

b. $W = +30.2$ L-atm?

Answer 5

a. -36 cal

b. $+736$ cal

OEIOU: $\# \text{cal} = \# \text{L-atm} \times \dfrac{\text{cal}}{\text{L-atm}}$

$= -1.5 \text{ L-atm} \times \dfrac{24.2 \text{ cal}}{1 \text{ L-atm}} = -36 \text{ cal}$

Question 6

In a certain chemical reaction, 10.3 kcal of heat are evolved. The volume of reactants is 36.8 liters; the volume of products is 9.2 liters. What is ΔE for the reaction?

Answer 6

-9.6 kcal

Defining equation: $\Delta E = Q - W$

OEIOU: $\Delta E = Q - p(V_2 - V_1)$ (# kcal/L-atm)

$= -10.3 \text{ kcal} - (1 \text{ atm})(9.2 \text{ L} - 36.8 \text{ L})$

$\times (0.0242 \text{ kcal/L-atm})$

$= -9.6 \text{ kcal}$

Statement 4

The only significant volume changes that occur during chemical reactions take place when gases are converted into **solid** or **liquid** products—which take up less space (volume)—or when solids or liquids are converted into gases and hence take up more space.

A mole of any gas occupies a volume of 24.4 liters at 25°C and 1 atm of pressure. Compared to this, the volume of a mole of liquid or solid may be considered to be negligible. (For example, a mole of liquid water occupies only 0.018 L.)

Question 7

For the following process at 25°C and 1 atm pressure, what is ΔV?

$$H_2O(l) + SO_3(g) \rightarrow H_2SO_4(l)$$

Answer 7

−24.4 liters

One mole of SO_3 gas was used up. Both H_2O and H_2SO_4 are *liquids* with relatively small volumes. Thus:

$$\Delta V = V_2 - V_1$$
$$= V_{H_2SO_4(l)} - (V_{H_2O(l)} + V_{SO_3(g)})$$
$$\approx 0\,L - (0\,L + 24.4\,L)$$
$$= -24.4\,L$$

Question 8

For the process:

$$H_2O(l) + SO_3(g) \rightarrow H_2SO_4(l),$$

$Q = -31.1$ kcal at 25°C and 1 atm. What is ΔE for the process?

Answer 8

−30.5 kcal

OEIOU: $\Delta E = Q - p(V_2 - V_1)$ (# kcal/L-atm)
$= -31.1$ kcal $- (1$ atm$)(-24.4$ L$)$
× (0.0242 kcal/L-atm)
$= -30.5$ kcal

PART 8—THE FIRST LAW OF THERMODYNAMICS

Question 9

For the process given below at 25°C and 1 atm, $Q = -189.1$ kcal. What is ΔE?

$$2O_2(g) + H_2S(g) \rightarrow H_2SO_4(l)$$

Answer 9

−187.3 kcal

OEIOU: $\Delta E = Q - p(V_2 - V_1)$ (# kcal/L-atm)

$= -189.1$ kcal $- (1$ atm$)$ $(0$ L $- 73.2$ L$)$

$\times$ $(0.0242$ kcal/L-atm$)$

$= -187.3$ kcal

Statement 5

The relationship between the heat energy or **enthalpy**, H, of a system and its **internal energy**, E, is given by the expression:

$$H = E + PV$$

where P and V are the pressure and volume of the system. Like E, H cannot be found, but the change in H, ΔH, is given by the equation:

$$\Delta H = \Delta E + \Delta(PV)$$

where $\Delta(PV)$ is the change in the product of P and V from the initial to the final state. If the pressure does not change:

$$(PV) = p\Delta V.$$

Question 10

ΔE for a certain process taking place at 1 atm pressure is -46.4 kcal. The volume of the system initially is 73.2 L; its final volume is 48.8 L. What is ΔH for the process?

Answer 10

−47.0 kcal

Defining equation: $\Delta H = \Delta E + \Delta(PV)$

OEIOU: $\Delta H = \Delta E + p(V_2 - V_1)$ (# kcal/L-atm)

Question 11

ΔH for a certain process taking place at 1 atm pressure is −53.0 kcal. The volume changes from 36.6 L to 24.4 L. What is ΔE for the process?

Answer 11

−52.7 kcal

OEIOU: $\Delta E = \Delta H - p(V_2 - V_1)$ (# kcal/L-atm)

Question 12

Consider the following process, which takes place at 25°C and 1 atm:

$$H_2(g) + I_2(s) \rightarrow 2HI(g)$$

If ΔE for the process is 13.0 kcal, what is ΔH?

Answer 12

13.6 kcal

Statement 6

Experimentally, ΔH may be found by measuring the heat flow for a process taking place at **constant pressure**. Mathematically,

$$\Delta H = Q_p$$

where the p subscript indicates that the pressure remains constant during the process. Heats of formation and reaction measured at constant pressure are **enthalpies** of formation and of reaction.

On the other hand, ΔE may be found experimentally by measuring Q for a process taking place at **constant volume**:

$$\Delta E = Q_v$$

Question 13

When 1.22 g of benzoic acid, $C_7H_6O_2$, is burned in a sealed bomb calorimeter at 25°C, the heat of reaction is −7.70 kcal. Is this ΔH_{rx} or ΔE_{rx}?

Answer 13

ΔE_{rx}

Since the reaction is carried out in a bomb, the volume is fixed. Therefore Q is $Q_v = \Delta E_{rx}$.

Question 14

To burn a certain organic compound in a bomb calorimeter at 25°C, the volume is fixed at 0.500 L, but the pressure changes from 2.00 atm to 1.00 atm. If ΔE for the reaction is −10.00 kcal, what is ΔH_{rx}?

Answer 14

−10.01 kcal

Defining equation: $\Delta H = \Delta E + \Delta (PV)$

OEIOU: $\Delta H = \Delta E + (P_2 V_2 - P_1 V_1)$ (# kcal/L-atm)

$= -10.00 \text{ kcal} + (1 \text{ atm} \times 0.500 \text{ L} - 2 \text{ atm}$
$\times 0.500 \text{ L})(0.0242 \text{ kcal/L-atm})$

$= -10.01 \text{ kcal}$

Question 15

The combustion (burning) of one mole of Mg metal in an open container at 25°C evolves 143.9 kcal of heat. Is this ΔH_{comb} or ΔE_{comb}?

Answer 15

ΔH_{comb}

Since the reaction is carried out in an open container, p = 1 atm throughout the reaction. Therefore Q is $Q_p = \Delta H$.

Question 16

What is ΔE for the combustion of 1.00 mole of Mg metal in an open container at 25°C and 1 atm, if ΔH_{comb} is −143.9 kcal?

Answer 16

−143.6 kcal

Reaction: $Mg(s) + \frac{1}{2} O_2(g) \rightarrow MgO(s)$

OEIOU: $\Delta E = \Delta H - p(V_2 - V_1)$ (# kcal/L-atm)

$= -143.9 \text{ kcal} - (1 \text{ atm})(-12.2 \text{ L})$

$\times (0.0242 \text{ kcal/L-atm})$

$= -143.6 \text{ kcal}$

Question 17

An organic compound is burned in a calorimeter. The heat flow accompanying the combustion is found to be −114 kcal/mole. Is this ΔH_{comb} or ΔE_{comb}?

Answer 17

This question cannot be answered on the basis of information supplied. We need to know whether the reaction was carried out in a bomb calorimeter (ΔE_{comb}) or in a calorimeter that would allow for constant pressure (ΔH_{comb}).

Statement 7

The breaking of a chemical bond requires energy; the formation of a bond releases energy. The energy absorbed in breaking Avogadro's number, one mole (6.023×10^{23}) of bonds of a given type, is called the bond dissociation energy—or simply **bond energy** (B.E.)—for that bond.

When all the bonds in a gaseous compound (at 25°C and 1 atm) are broken, and the compound is thereby reconverted into its constituent atoms, the enthalpy change is called the **heat of atomization**, ΔH_{atom}. When the compound is a simple diatomic molecule, this enthalpy change is often called the **heat of dissociation**, ΔH_{diss}, and is the same as bond energy.

It is convenient to use Hess's Law and data such as B.E., ΔH_{atom}, and ΔH_{diss}, which have been determined and can be found in tables, to calculate ΔH_{rx} and other enthalpy changes.

For example, for the generalized process: $AB + CD \rightarrow AD + CB \qquad \Delta H_{rx} = ?$

$AB \rightarrow A + B \qquad \Delta H_1 = \Delta H_{diss, AB}$

$CD \rightarrow C + D \qquad \Delta H_2 = \Delta H_{diss, CD}$

PART 8—THE FIRST LAW OF THERMODYNAMICS

$$A + D \rightarrow AD \qquad \Delta H_3 = -\Delta H_{diss, AD}$$
$$C + B \rightarrow CB \qquad \Delta H_4 = -\Delta H_{diss, CB}$$

Sum: $AB + CD \rightarrow AD + CB$

$$\Delta H_{rx} = \Delta H_{diss, AB} + \Delta H_{diss, CD} - \Delta H_{diss, AD} - \Delta H_{diss, CB}$$

Question 18

Given these values:

$\Delta H_{diss, HCl}$ = 103 kcal/mole

$\Delta H_{diss, H_2}$ = 104 kcal/mole

$\Delta H_{diss, Cl_2}$ = 58 kcal/mole

What is ΔH_{rx} for:

$$H_2 + Cl_2 \rightarrow 2HCl$$

Answer 18

−44 kcal

$$H_2 \rightarrow 2H \qquad \Delta H_1 = \Delta H_{diss, H_2}$$
$$Cl_2 \rightarrow 2Cl \qquad \Delta H_2 = \Delta H_{diss, Cl_2}$$
$$2H + 2Cl \rightarrow 2HCl \qquad \Delta H_3 = 2(-\Delta H_{diss, HCl})$$

Sum: $H_2 + Cl_2 \rightarrow 2HCl$

$$\Delta H_{rx} = \Delta H_1 + \Delta H_2 + \Delta H_3$$
$$= \Delta H_{diss, H_2} + \Delta H_{diss, Cl_2} + 2(-\Delta H_{diss, HCl})$$
$$= 104 \text{ kcal} + 58 \text{ kcal} + 2(-103 \text{ kcal})$$
$$= -44 \text{ kcal}$$

Question 19

Given:

$\Delta H_{atom, H_2O}$ = 221 kcal/mole

$\Delta H_{diss, H_2}$ = 104 kcal/mole

$\Delta H_{diss, O_2}$ = 118 kcal/mole

What is ΔH_{rx} for:

$$2H_2O \rightarrow 2H_2 + O_2$$

Answer 19

116 kcal

$$\begin{array}{ll} 2(H_2O \rightarrow 2H + O) & \Delta H_1 = 2\Delta H_{atom,H_2O} \\ 2(2H \rightarrow H_2) & \Delta H_2 = -\Delta H_{diss,H_2} \\ 2O \rightarrow O_2 & \Delta H_3 = -\Delta H_{diss,O_2} \end{array}$$

Sum: $2H_2O \rightarrow 2H_2 + O_2$

$$\Delta H_{rx} = 2\Delta H_1 + 2\Delta H_2 + \Delta H_3$$
$$= 2(221 \text{ kcal}) + 2(-104 \text{ kcal}) + (-118 \text{ kcal})$$
$$= 116 \text{ kcal}$$

Statement 8

An alternative approach to using the preceding types of relationships to calculate ΔH_{rx} is to employ this simple relationship:

$$\Delta H_{rx} = (B.E.)_{broken} - (B.E.)_{formed}$$

Question 20

Given:

$(B.E.)_{H-H} = 104$ kcal/mole

$(B.E.)_{Br-Br} = 46$ kcal/mole

$(B.E.)_{H-Br} = 88$ kcal/mole

What is ΔH_{rx} for the formation of two moles of HBr from H_2 and Br_2?

Answer 20

−26 kcal

Reaction: $H_2 + Br_2 \rightarrow 2HBr$

$$\Delta H_{rx} = (B.E.)_{broken} - (B.E.)_{formed}$$

$$= (B.E._{H_2} + B.E._{Br_2}) - 2 B.E._{HBr}$$
$$= (104 \text{ kcal} + 46 \text{ kcal}) - 2(88 \text{ kcal})$$
$$= -26 \text{ kcal}$$

Question 21

If the B.E. for N≡N is 226 kcal/mole, the B.E. for H-H is 104 kcal/mole, and the B.E. for N-H is 93 kcal/mole, what is ΔH_{rx} when NH_3 is formed from N_2 and H_2?

Answer 21

-20 kcal

Reaction: $H_2 + 3H_2 \rightarrow 2NH_3$

$$\Delta H_{rx} = (B.E._{N_2} + 3 B.E._{H_2}) + 6 B.E._{NH}$$
$$= (226 \text{ kcal}) + 3(10 \text{ kcal}) - 6(93 \text{ kcal})$$
$$= -20 \text{ kcal}$$

Question 22

Given:

$(B.E.)_{H-O} = 111$ kcal/mole

$(B.E.)_{H-H} = 104$ kcal/mole

$(B.E.)_{O_2} = 118$ kcal/mole

What is the heat of reaction for:

$$2H_2O \rightarrow 2H_2 + O_2$$

Answer 22

118 kcal

$$\Delta H_{rx} = 4 B.E._{HO} - (2 B.E._{H_2} + B.E._{O_2})$$
$$= 4(111 \text{ kcal}) - 2(104 \text{ kcal}) - 118 \text{ kcal}$$
$$= 118 \text{ kcal.}$$

(Compare with Answer 19.)

TESTING YOUR MASTERY—Part 8
The First Law of Thermodynamics

Question 23

When one mole of liquid Br_2 is converted to Br_2 vapor at 25°C and 1 atm pressure, 7.3 kcal of heat is absorbed and 0.59 kcal of expansion work is done by the system. What is ΔE for this vaporization process?

Answer 23

+6.7 kcal

OEIOU: $\Delta E = Q - W$

Question 24

What is W (work) for the following process at 25°C and 1.00 atm, expressed a) in L-atm; and b) in kcal:

$$N_2O_4(g) \rightarrow 2NO_2(g)$$

Answer 24

a. 24.4 L-atm

OEIOU: $W = p(V_{NO_2} - V_{N_2O_4})$

b. 0.590 kcal

OEIOU: $W = p(V_{NO_2} - V_{N_2O_4}) \times \frac{kcal}{L\text{-}atm}$

Question 25

Express W in L-atm and kcal for the following reaction at 25°C and 1.00 atm:

$$3Fe(s) + 2O_2(g) \rightarrow Fe_3O_4(s)$$

Answer 25

−48.8 L-atm or −1.18 kcal

OEIOU: $W = p(V_{Fe_3O_4} - V_{3Fe} - V_{2O_2})$

or: $W \approx -pV_{2O_2}$

Question 26

What is W at 25°C and 1.00 atm when 25.4 g of I_2 is sublimed (converted from solid to vapor)?

Answer 26

+2.44 L-atm or +0.0590 kcal

OEIOU: $W = p(V_{I_2,gas} - V_{I_2,solid}) \times \dfrac{\#g\ I_2}{GMW_{I_2}}$

or: $W \approx pV_{I_2,gas} \times \dfrac{\#g\ I_2}{GMW_{I_2}}$

Question 27

For the sublimation of I_2 at 25°C and 1.00 atm:

$$I_2(s) \rightarrow I_2(g)$$

Q is +14.9 kcal/mole. What is ΔE for the sublimation process?

Answer 27

+14.3 kcal

OEIOU: $\Delta E \approx Q - (pV_{I_2,gas})\left(\dfrac{kcal}{L\text{-atm}}\right)$

Question 28

When one mole of graphite (carbon) is burned at 25°C and 1.00 atm, $Q = -94.1$ kcal. What is ΔE for this combustion?

Answer 28

−94.1 kcal

OEIOU: $\Delta E = Q - p(V_{CO_2} - V_{O_2} - V_C)\left(\dfrac{kcal}{L\text{-atm}}\right)$

or: $\Delta E \approx Q$

Question 29

The volume of a certain system undergoes a change from 12.6 L to 37.8 L during a chemical reaction taking place at 1.00 atm. If ΔE is +11.6 kcal, what is ΔH for the process?

Answer 29

$\Delta H = 12.2$ kcal

OEIOU: $\Delta H = \Delta E + p(V_2 - V_1) \left(\dfrac{\text{kcal}}{\text{L-atm}}\right)$

Question 30

ΔE for the process $2CO(g) + O_2(g) \rightarrow 2CO_2(g)$ is -134.8 kcal at 25°C and 1.00 atm. Calculate ΔH for this oxidation of CO to CO_2.

Answer 30

$\Delta H = -135.4$ kcal

OEIOU: $\Delta H + p(V_{2CO_2} - V_{2CO} - V_{O_2}) \left(\dfrac{\text{kcal}}{\text{L-atm}}\right)$

$\Delta H = \Delta E - pV_{O_2} \times \left(\dfrac{\text{kcal}}{\text{L-atm}}\right)$

Question 31

Q has different values for the formation of water as a gas and as a liquid in an open container at 25°C and 1.00 atm:

$2H_2(g) + O_2(g) \rightarrow 2H_2O(g)$ $Q_1 = -115.6$ kcal

$2H_2(g) + O_2(g) \rightarrow 2H_2O(l)$ $Q_2 = -136.6$ kcal

a. What is ΔH for each of these processes?

b. What is ΔE for each of these processes?

Answer 31

a. $\Delta H_1 = -115.6$ kcal

$\Delta H_2 = -136.6$ kcal

OEIOU: $\Delta H = Q_p$

b. $\Delta E_1 = -115.0$ kcal

$\Delta E_2 \approx -134.8$ kcal

OEIOU: $\Delta E_1 = \Delta H_1 - pV_{O_2} \left(\dfrac{\text{kcal}}{\text{L-atm}}\right)$

OEIOU: $\Delta E_2 \approx \Delta H_2 - p\left(-V_{2H_2} - V_{O_2}\right)\left(\dfrac{\text{kcal}}{\text{L-atm}}\right)$

Question 32

Q is -17.3 kcal for the formation of a mole of CH_4 gas from graphite (carbon) and H_2 gas at 25°C in a sealed container.

a. What is ΔH_{rx}?

b. What is ΔE_{rx}?

Answer 32

a. -17.9 kcal

OEIOU: $\Delta H_{rx} \approx Q_v + p(V_{CH_4} - V_{2H_2O})$

b. -17.3 kcal

OEIOU: $\Delta E_{rx} = Q_v$

Question 33

Given the following values:

$\Delta H_{\text{diss},N_2} = 226$ kcal/mole;

$\Delta H_{\text{diss},O_2} = 118$ kcal/mole;

$\Delta H_{\text{diss},NO} = 297$ kcal/mole;

what is ΔH_{rx} for the following reaction:

$N_2 + O_2 \rightarrow 2NO$

Answer 33

−250 kcal

OEIOU: $\Delta H_{rx} = \Delta H_{diss,N_2} + \Delta H_{diss,O_2} - 2\Delta H_{diss,NO}$

Question 34

Given the following values:

$\Delta H_{atom,C}$ = 171 kcal/mole;

$\Delta H_{diss, Cl_2}$ = 28.9 kcal/mole;

and that ΔH_{rx} = −27.0 kcal for:

$C(s) + 2Cl_2(g) \rightarrow CCl_4(g)$;

calculate ΔH_{atom} for $CCl_4(g)$.

Answer 34

+256 kcal/mole

OEIOU: $\Delta H_{atom,CCl_4} = \Delta H_{atom,C} + 2\Delta H_{diss, Cl_2} - \Delta H_{rx}$

Question 35

Given these values:

$(B.E.)_{C-H}$ = 99 kcal/mole

$(B.E.)_{Cl-Cl}$ = 58 kcal/mole

$(B.E.)_{C-Cl}$ = 79 kcal/mole

$(B.E.)_{H-Cl}$ = 103 kcal/mole

Calculate ΔH_{rx} for:

$CH_4(g) + 4Cl_2(g) \rightarrow CCl_4(g) + 4HCl(g)$

Answer 35

−100 kcal

OEIOU: ΔH_{rx} = 4B.E.$_{CH}$ + 4B.E.$_{Cl_2}$ − 4B.E.$_{CCl}$ − 4B.E.$_{HCl}$

Question 36

If B.E. for N_2 is 226 kcal/mole; B.E. for Cl_2 is 58 kcal/mole; and ΔH_{rx} is +55.0 kcal for the reaction $N_2(g) + 3Cl_2(g) \rightarrow 2NCl_3(g)$; what is B.E. for N–Cl calculated from these data?

Answer 36

+58 kcal/mole

OEIOU: $\text{B.E.}_{NCl} = \frac{1}{6}(\text{B.E.}_{N_2} + 3\text{B.E.}_{Cl_2} - \Delta H_{rx})$

PART 9
THE SECOND LAW OF THERMODYNAMICS

Statement 1

The **entropy** of the universe is increasing. This is a statement of the second law of thermodynamics. Entropy, S, is a measure of the **disorder** or **freedom** or **randomness** of a system; the more disorganized the system, the greater the positive value of S. We are often interested in finding ΔS, the change in entropy, accompanying a change in state. For the statement

$$\text{System (state 1)} \rightarrow \text{System (state 2)}$$

if state 2 is **more disorganized** (greater freedom) than state 1, ΔS will be **positive**;

if state 2 is **less disorganized** (less freedom) than state 1, ΔS will be **negative**.

Question 1

Will ΔS be positive or negative for the following process:

$$H_2O(l) \rightarrow H_2O(g)$$

> **Answer 1**
>
> positive
>
> H_2O as a gas has greater freedom—is more disorganized—than liquid H_2O.

Question 2

What will be the sign on ΔS for the following process:

$$H_2(g) + Cl_2(g) \rightarrow 2HCl(g)$$

> **Answer 2**
>
> positive
>
> The distribution of H and Cl is less random when they are held to identical atoms in H_2 and Cl_2.

Question 3

Will ΔS be positive or negative for the following reaction?

$$2H_2(g) + N_2(g) \rightarrow N_2H_4(l)$$

Answer 3

negative

Statement 2

Unlike internal energy values, E, and enthalpy values, H, it is possible to determine absolute values, S°, for the entropy of a system at 25°C and 1 atm. Knowing these, the entropy of reaction, S°_{rx}, can be found:

$$\Delta S^\circ_{rx} = S^\circ_{products} - S^\circ_{reactants}$$

Question 4

Given:

S°_C = 1.36 cal/mole °K

$S^\circ_{H_2}$ = 31.2 cal/mole °K

$S^\circ_{CH_4}$ = 44.5 cal/mole °K

What is ΔS°_{rx} for: $C + 2H_2 \rightarrow CH_4$?

Answer 4

−19.3 cal/°K

$$\Delta S^\circ_{rx} = S^\circ_{prod} - S^\circ_{react}$$
$$= S^\circ_{CH_4} - (S^\circ_C + 2S^\circ_{H_2})$$
$$= (44.5) - (1.36) - 2(31.2)$$
$$= -19.3 \text{ cal/°K}$$

Note: The mole units are cancelled by multiplying (not shown here) by the number of moles of CH_4, C, and H_2 involved.

Question 5

Calculate ΔS°_{rx} for the process

$$C_2H_4 + H_2 \rightarrow C_2H_6$$

given the following data:

$S^\circ_{C_2H_4}$ = 52.4 cal/mole °K

$S^\circ_{H_2}$ = 31.2 cal/mole °K

$S^\circ_{C_2H_6}$ = 54.8 cal/mole °K

Answer 5

-28.8 cal/°K

Statement 3

The work accompanying a chemical process may be either **unavoidable work** (unavoidable because products take up more or less volume than reactants) or **useful work**.

Every system has a capacity to do useful work, which is called **Gibbs free energy**, G.

As with enthalpy, H, and internal energy, E, absolute values of G cannot be found, but ΔG values for changes of state can be found. For a given process at 25°C and 1 atm, we use the symbol ΔG°_{rx}, and

$$\Delta G^\circ_{rx} = \Delta G^\circ_{f,\,products} - \Delta G^\circ_{f,\,reactants}$$

where ΔG°_f represents the **standard free energy of formation** of compounds (products or reactants) from their elements. The standard free energy of formation of elements is zero.

Question 6

Given:

$\Delta G^\circ_{f,\,CH_4}$ = -12.1 kcal/mole

$\Delta G^\circ_{f,\,CO_2}$ = -94.3 kcal/mole

$\Delta G^\circ_{f,\,H_2O}$ = -56.7 kcal/mole

What is ΔG°_{rx} for the following reaction:

$$CH_4 + 2O_2 \rightarrow 2H_2O + CO_2$$

Answer 6

−195.6 kcal

$$\Delta G^\circ_{rx} = \Delta G^\circ_{f, \text{prod}} - \Delta G^\circ_{f, \text{react}}$$

$$= (2\Delta G^\circ_{f, H_2O} + \Delta G^\circ_{f, CO_2}) - (\Delta G^\circ_{f, CH_4} + 2\Delta G^\circ_{f, O_2})$$

$$= 2(-56.7 \text{ kcal}) + (-94.3 \text{ kcal}) - (-12.1 \text{ kcal}) - 2(0)$$

$$= -195.6 \text{ kcal}$$

Question 7

Given:

$\Delta G^\circ_{f, HCl} = -22.8$ kcal/mole

$\Delta G^\circ_{f, H_2O} = -56.7$ kcal/mole

What is ΔG°_{rx} for the following process:

$$2H_2O + 2Cl_2 \rightarrow 4HCl + O_2$$

Answer 7

+22.2 kcal

$$\Delta G^\circ_{rx} = (4\Delta G^\circ_{f, HCl} + \Delta G^\circ_{f, O_2}) - (2\Delta G^\circ_{f, H_2O} + \Delta G^\circ_{f, Cl_2})$$

$$= 4(-22.8 \text{ kcal}) + (0) - 2(-56.7 \text{ kcal}) - (0)$$

$$= +22.2 \text{ kcal}$$

Statement 4

A reaction is **spontaneous** (i.e., it proceeds "automatically") only if the capacity of the system to do work decreases during the process. In terms of ΔG,

if ΔG is **negative**, the process is **spontaneous**;
if ΔG is **positive**, the process is **nonspontaneous**.

Question 8

Given: $\Delta G^\circ_{ICl} = -1.3$ kcal/mole

Would the following reaction be spontaneous at 25°C and 1 atm?

$$I_2 + Cl_2 \rightarrow 2ICl$$

Answer 8

Yes; ΔG_{rx}° is negative.

$$\Delta G_{rx}^{\circ} = 2\Delta G_{f,\,ICl}^{\circ} - \Delta G_{f,\,I_2}^{\circ} - \Delta G_{f,\,Cl_2}^{\circ}$$
$$= 2(-1.3 \text{ kcal}) - (0) - (0)$$
$$= -2.6 \text{ kcal}$$

Question 9

If the standard free energy of formation of HgO is -14.0 kcal/mole, would the following reaction be spontaneous?

$$2HgO \rightarrow 2Hg + O_2$$

Answer 9

No; ΔG_{rx}° is positive (+28.0 kcal).

Statement 5

The value of ΔG, and thus the spontaneity of a reaction is determined by two factors:

ΔH the more **negative** the enthalpy change, the greater the probability that the process will be spontaneous; and

ΔS the more **positive** the entropy, the greater the probability that the process will be spontaneous.

The precise relationship of ΔG, ΔH, and ΔS is given by the **Gibbs-Helmholtz equation**:

$$\Delta G = \Delta H - T\Delta S$$

Question 10

Given: $\Delta H_{f,\,HBr}^{\circ} = -8.7$ kcal/mole.

If ΔS° is +13.7 cal/mole °K for the process

$$H_2 + Br_2 \rightarrow 2HBr$$

what is ΔG_{rx}° at 25°C?

Answer 10

-25.6 kcal

$$\Delta G^\circ_{rx} = \Delta H^\circ_{rx} - T\Delta S^\circ_{rx}$$
$$= 2\Delta H^\circ_{f, HBr} - T \times 2\Delta S^\circ$$
$$= 2(-8.7 \text{ kcal}) - (298\ ^\circ K)(2)(0.0137 \text{ kcal/mole } ^\circ K)$$
$$= -25.6 \text{ kcal}$$

Note: Change °C to °K, and cal to kcal.

Question 11

Predict whether the process $2CO + O_2 \rightarrow 2CO_2$ is spontaneous at 25°C and 1 atm pressure, given:

$\Delta H^\circ_{f, CO} = -26.4$ kcal/mole

$\Delta H^\circ_{f, CO_2} = -94.0$ kcal/mole

$S^\circ_{CO} = 47.3$ cal/mole °K

$S^\circ_{O_2} = 49.0$ cal/mole °K

$S^\circ_{CO_2} = 51.1$ cal/mole °K

Answer 11

Yes, it would be spontaneous; ΔG is negative.

$$\Delta G^\circ_{rx} = (2\Delta H^\circ_{f, CO_2} - 2\Delta H^\circ_{f, CO} - \Delta H^\circ_{f, O_2}) - T(2S^\circ_{CO_2} - 2S^\circ_{CO} - S^\circ_{O_2})$$
$$= -122.9 \text{ kcal}$$

Question 12

Predict whether the process $2Ag_2O \rightarrow 4Ag + O_2$ is spontaneous at 25°C and 1 atm, given:

$\Delta H^\circ_{f, Ag_2O} = -7.3$ kcal/mole

$\Delta S^\circ_{rx} = +31.6$ cal/°K

Answer 12

Nonspontaneous; ΔG is positive.

$$\Delta G^\circ_{rx} = \Delta H^\circ_{rx} - T\Delta S^\circ_{rx}$$

$$= 4\Delta H^\circ_{f, Ag} + \Delta H^\circ_{f, O_2} - 2\Delta H^\circ_{f, Ag_2O} - T\Delta S^\circ_{rx}$$

$$= 5.2 \text{ kcal}$$

Question 13

What is ΔS°_{rx} at 25°C for the process $N_2 + 3H_2 \rightarrow 2NH_3$, given:

$\Delta H^\circ_{rx} = -22.0$ kcal

$\Delta G^\circ_{rx} = -8.0$ kcal

Answer 13

-46.9 cal/°K

$$\Delta G^\circ_{rx} = \Delta H^\circ_{rx} - T\Delta S^\circ_{rx}$$

OEIOU: $\Delta S^\circ_{rx} = \dfrac{\Delta H^\circ_{rx} - \Delta G^\circ_{rx}}{T}$

$$= \frac{-22.0 \text{ kcal} - (-8.0 \text{ kcal})}{298°K}$$

$$= -0.0469 \text{ kcal/°K}$$

or -46.9 cal/°K

Question 14

ΔG is +49.8 kcal at 25°C for the reaction $2N_2 + O_2 \rightarrow 2N_2O$.

If ΔH is 39.0 kcal and ΔS is -36.2 cal/mole °K for the reaction:

a. What is ΔG at 900 °C?

b. What effect does increasing the temperature have on the spontaneity of the reaction?

Answer 14

a. +81.5 kcal

OEIOU: $\Delta G = \Delta H - T\Delta S$

b. The likelihood of N_2O forming spontaneously from N_2 and O_2 at 900°C is less than that at 25°C. Had the ΔS term been *positive* rather than *negative*, spontaneity could have resulted at a sufficiently high temperature.

TESTING YOUR MASTERY—Part 9
The Second Law of Thermodynamics

Question 15

For which of the following processes would you expect ΔS to be positive?

a. $I_2(g) \rightarrow I_2(s)$

b. $H_2(g) + O_2(g) \rightarrow H_2O_2(l)$

c. $2KClO_3(s) \rightarrow 2KCl(s) + 3O_2(g)$

d. $P_4(s) \rightarrow 4P(g)$

Answer 15

ΔS is positive for (c) and (d) only.

Question 16

Given:

$S^°_{H_2(g)} = +31.2$ cal/mole °K

$S^°_{O_2(g)} = +49.0$ cal/mole °K

$S^°_{H_2O_2(l)} = +54.0$ cal/mole °K

What is $\Delta S^°_{rx}$ for the following process:

$$H_2(g) + O_2(g) \rightarrow H_2O_2(l)$$

Answer 16

-26.2 cal/mole $°K$

OEIOU: $\Delta S°_{rx} = S°_{H_2O_2} - S°_{H_2} - S°_{O_2}$

Question 17

Given the $S°$ values for $KClO_3(s)$, $KCl(s)$, and $O_2(g)$ of $+34.2$ cal/mole$°K$, $+19.8$ cal/mole$°K$, and $+49.0$ cal/mole$°K$, respectively, calculate $\Delta S°_{rx}$ for:

$$2KClO_3(s) \rightarrow 2KCl(s) + 3O_2(g)$$

Answer 17

$+118.2$ cal/mole$°K$

OEIOU: $\Delta S°_{rx} = 2S°_{KCl} + 3S°_{O_2} + 3S°_{KClO_3}$

Question 18

Given these values:

$\Delta G°_{f, CH_4(g)} = -12.1$ kcal/mole,

$\Delta G°_{f, CCl_4(g)} = -15.3$ kcal/mole,

$\Delta G°_{f, HCl(g)} = -22.8$ kcal/mole,

What is $\Delta G°_{rx}$ for:

$$CH_4(g) + 4Cl_2(g) \rightarrow CCl_4(g) + 4HCl(g)$$

Answer 18

-94.4 kcal

OEIOU: $\Delta G°_{rx} = \Delta G°_{f, CCl_4} + 4\Delta G°_{f, HCl} - \Delta G°_{f, CH_4} - \Delta G°_{f, Cl_2}$

Question 19

Would the reaction in either Question 16 or Question 17 be spontaneous?

PART 9-THE SECOND LAW OF THERMODYNAMICS   185

Answer 19

Yes. The reaction

$$CH_4 + 4Cl_2 \rightarrow CCl_4 + 4HCl$$

would be spontaneous, since ΔG is negative.

Question 20

If $\Delta G^\circ_{f, MgO}(s)$ is -136.1 kcal/mole, and $\Delta G^\circ_{f, H_2O}(l)$ is -56.7 kcal/mole, would the following be a spontaneous reaction at 25°C and 1 atm pressure?

$$MgO(s) + H_2(g) \rightarrow H_2O(l) + Mg(s)$$

Answer 20

No. ΔG_{rx} is positive (+79.4 kcal).

OEIOU: $\Delta G^\circ_{rx} = \Delta G^\circ_{f, H_2O} + \Delta G^\circ_{f, Mg} - \Delta G^\circ_{f, MgO} - \Delta G^\circ_{f, H_2}$

Question 21

The decomposition of solid $CaCO_3$ into solid CaO and CO_2 gas at 25°C is endothermic by 42.5 kcal/mole. ΔS° for the process is +38.6 cal/mole°K. What is ΔG° for this decomposition?

Answer 21

+31.0 kcal

OEIOU: $\Delta G^\circ = \Delta H^\circ - T\Delta S^\circ$

Question 22

The process $NH_3(g) + HCl(g) \rightarrow NH_4Cl(s)$ is exothermic by -42.1 kcal at 25°C and 1.00 atm.

a. Find ΔG°_{rx}, given $S^\circ_{NH_3} = +46.0$ cal/mole °K; $S^\circ_{HCl} = +44.6$ cal/mole °K; and $S^\circ_{NH_4Cl} = +22.6$ cal/mole °K.

b. Would the reaction be spontaneous?

Answer 22

−21.8 kcal

OEIOU: $\Delta G_{rx}^\circ = \Delta H_f^\circ - T(S_{NH_4Cl}^\circ - S_{NH_3}^\circ - S_{HCl}^\circ)$

b. Yes. ΔG_{rx} is negative.

Question 23

Given the values

$\Delta H_{rx}^\circ = -39.0$ kcal

$\Delta G_{rx}^\circ = -49.8$ kcal

find ΔS_{rx}° for the following process:

$$2N_2O(g) \rightarrow 2N_2(g) + O_2(g)$$

Answer 23

+36.2 cal/mole °K

OEIOU: $\Delta S_{rx}^\circ = (\Delta H_{rx}^\circ - \Delta G_{rx}^\circ) / T$

PART 10
GASES

Statement 1

In the **gaseous state** a substance expands to fill all available space. The average distance between molecules or atoms comprising a gas is great, making it easy to compress the gas, and making the action of the constituent molecules or atoms very sensitive to changes in temperature. These pressure and temperature effects are dealt with quantitatively by "gas laws," which are discussed in this section.

The **pressure** of a gas is usually measured with a manometer calibrated in **torrs** (or **mm Hg**), the height that a column of mercury is raised by the imbalance between the pressure of the gas in question and a standard pressure (usually the atmosphere). Gas laws, on the other hand, often require expression of pressures in terms of "atmospheres": 1 atm = 760 mm Hg.

The **temperature** of a gas is usually measured with a thermometer calibrated in **degrees Celsius** (or centigrade), whereas gas laws require expression of absolute temperature measured in **degrees Kelvin**: $°K = °C + 273°$.

Question 1

Convert these pressures to atm units:

a. 1140 mm Hg

b. 456 torr

Answer 1

a. 1.50 atm

b. 0.600 atm

$$\#a = \#b \times C.F.$$

OEIOU: $\#\text{atm} = \#\text{mm Hg} \times \dfrac{\text{atm}}{\text{mm Hg}}$

Question 2

Convert these temperatures to $°K$:

a. 0°C

b. 25°C

Answer 2

a. 273°K

b. 298°K

OEIOU: °K = °C + 273°

Statement 2

Boyle's Law states that the volume of a gas varies **inversely** with its pressure, provided that the temperature does not change. Mathematically:

$$V_T = \frac{k_T}{p}$$

where the subscript T means that temperature is held constant, k_T is a proportionality constant, and p is the pressure.

Question 3

A certain gas occupies 155 ml at 25°C. If the confining pressure is changed from 760 mm Hg to 1520 mm, what will be the new volume taken up by the gas?

Answer 3

77.5 ml

$$V_1 = \frac{k_T}{p_1} \text{ or } k_T = V_1 p_1$$

$$V_2 = \frac{k_T}{p_2} \text{ or } k_T = V_2 p_2$$

Thus: $V_1 p_1 = V_2 p_2$

OEIOU: $V_2 = V_1 \times \frac{p_1}{p_2}$

$$= 155 \text{ ml} \times \frac{760 \text{ mm}}{1520 \text{ mm}}$$

$$= 77.5 \text{ ml}$$

Question 4

The pressure on 155 ml of oxygen at 0°C is changed from 1.00 atm to 0.500 atm. What volume will the gas occupy after the pressure change?

> ### Answer 4
>
> 310 ml
>
> $$V_1 P_1 = V_2 P_2$$
>
> OEIOU: $V_2 = V_1 \times p_1/p_2$
>
> $$= 155 \text{ ml} \times \frac{1.00 \text{ atm}}{0.500 \text{ atm}}$$
>
> $$= 310 \text{ ml}$$

Question 5

A sample of nitrogen occupies 155 ml at 30°C and 760 torr. To what must the pressure be changed if the gas is to occupy 200 ml?

> ### Answer 5
>
> 589 torr
>
> $$V_1 P_1 = V_2 P_2$$
>
> OEIOU: $p_2 = \dfrac{V_1}{V_2} \times p_1$
>
> $$= \frac{155 \text{ ml}}{200 \text{ ml}} \times 760 \text{ torr}$$
>
> $$= 589 \text{ torr}$$

Statement 3

Charles' Law states that the volume of a gas varies **directly** with its **absolute temperature**, provided that the pressure does not change. Mathematically:

$$V_p = k_p T$$

where the subscript p means that the pressure is held constant, k_p is a proportionality constant, and T is the temperature in degrees Kelvin.

Question 6

A sample of acetylene occupies 1.00 L at 273°K and 1 atm pressure. If the temperature is raised to 341°K, what will be the new volume?

Answer 6

1.25 L

$V_1 = k_p T_1$ or $k_p = V_1/T_1$

$V_2 = k_p T_2$ or $k_p = V_2/T_2$

Thus: $V_1/T_1 = V_2/T_2$

OEIOU: $V_2 = V_1 \times T_2/T_1$

$= 1.00 \text{ L} \times \dfrac{341°K}{273°K}$

$= 1.25 \text{ L}$

Question 7

What volume will a 1.00-L sample of chlorine occupy at 20°C and 1 atm if the temperature is increased to 40°C?

Answer 7

1.07 L

$V_1/T_1 = V_2/T_2$

OEIOU: $V_2 = V_1 \times T_2/T_1$

$= 1.00 \text{ L} \times \dfrac{313°K}{293°K}$

$= 1.07 \text{ L}$

Question 8

A sample of methane occupies 1.00 L at 0°C and 1 atm. How much must the temperature be changed in order to increase the volume to 2.00 L?

Answer 8

To $273°$ C (or $546°$ K)

$V_1/T_1 = V_2/T_2$

OEIOU: $T_2 = V_2/V_1 \times T_1$

$= \dfrac{2.00 \text{ L}}{1.00 \text{ L}} \times 273°\text{K}$

$= 546° \text{ K or } 273°\text{C}$

Statement 4

The **ideal gas equation** combines the generalizations made in Boyle's Law and Charles' Law, plus the fact that the volume occupied by a gas is directly proportional to the number of moles of the gas. It is usually written as follows:

$$pV = nRT$$

where n is the number of moles, and R is a proportionality constant called the "gas constant": $R = 0.0821$ L-atm/mole°K.

Question 9

What volume does exactly one mole of helium occupy at $0°$C and 1.00 atm?

Answer 9

22.4 L

$pV = nRT$

OEIOU: $V = nRT/p$

$= \dfrac{(1.00 \text{ mole}) (0.0821 \text{ L-atm/mole}°\text{K}) (273°\text{K})}{1.00 \text{ atm}}$

$= 22.4$ L

Question 10

What volume would 30.2 g of neon occupy at $10°$C and 740 mm pressure?

Answer 10

35.7 L

$pV = nRT$

$= \dfrac{\#g}{GAW_{Ne}} \times RT$

OEIOU: $V = \dfrac{\#g}{GAW_{Ne}} \times \dfrac{RT}{p}$

$= \dfrac{30.2 \text{ g}}{20.2 \text{ g/mole}} \times \dfrac{(0.0821 \text{ L-atm/mole}°\text{K})(283°\text{K})}{0.974 \text{ atm}}$

$= 35.7 \text{ L}$

Question 11

At what pressure would one mole of helium at 0°C occupy 16.8 liters?

Answer 11

1.33 atm

$pV = nRT$

OEIOU: $p = \dfrac{nRT}{V}$

$= \dfrac{(1.00 \text{ mole})(0.0821 \text{ L-atm/mole}°\text{K})(273°\text{K})}{16.8 \text{ L}}$

$= 1.33 \text{ atm}$

Question 12

What will be the new volume of 4.00 L of argon if the pressure and temperature are changed from 1.00 atm and 293°K to 3.00 atm and 439°K?

Answer 12

2.00 L

$p_1 V_1 = nRT$ or $nR = p_1 V_1 / T_1$

$p_2 V_2 = nRT_2$ or $nR = p_2 V_2 / T_2$

Thus: $V_2 = \dfrac{p_1}{p_2} \times \dfrac{T_2}{T_1} \times V_1$

$= \dfrac{1.00 \text{ atm}}{3.00 \text{ atm}} \times \dfrac{439°K}{293°K} \times 4.00 \text{ L}$

$= 2.00 \text{ L}$

Question 13

A sample of hydrogen occupies 515 ml at 20°C and 760 mm. If the temperature is changed to 0°C and the pressure to 660 mm, what will be the new volume of the gas?

Answer 13

553 ml

$\dfrac{p_1 V_1}{T_1} = \dfrac{p_2 V_2}{T_2}$

OEIOU: $V_2 = \dfrac{p_1}{p_2} \times \dfrac{T_2}{T_1} \times V_1$

$= \dfrac{760 \text{ mm}}{660 \text{ mm}} \times \dfrac{273°K}{293°K} \times 515 \text{ ml}$

$= 553 \text{ ml}$

Question 14

A 2.00-g sample of helium occupies 11.2 L at 0°C and 1.00 atm. If the pressure is doubled, to what must the temperature be changed to keep the volume from changing?

Answer 14

273°C (or 546°K)

$\dfrac{p_1 V_1}{T_1} = \dfrac{p_2 V_2}{T_2}$

But $V_1 = V_2$, so: $\dfrac{p_1}{T_1} = \dfrac{p_2}{T_2}$

OEIOU: $T_2 = T_1 \times \dfrac{p_2}{p_1}$

$= 273° \text{ K} \times \dfrac{2.00 \text{ atm}}{1.00 \text{ atm}}$

$= 546° \text{ K or } 273°C$

Statement 5

The **vapor density method** makes use of the ideal gas equation to determine the molecular weight of a gas or volatile liquid. The liquid is heated to convert it into a vapor (gas). Its temperature and pressure are then determined, and the molecular weight is calculated, making use of the relationship

$$n = \frac{\#g}{GMW}$$

Question 15

A 0.412-g sample of chloroform is collected as 100.0 ml of a gas in a flask. At 75°C the pressure of the vapor is 741 mm. Calculate the molecular weight of chloroform.

Answer 15

121 g/mole

$$pV = nRT$$

$$= \frac{\#g}{GMW} \times RT$$

OEIOU: $GMW = \#g \times \dfrac{RT}{pV}$

$$= \frac{(0.412 \text{ g})(0.0821 \text{ L-atm/mole}°K)(348°K)}{(0.975 \text{ atm})(0.100 \text{ L})}$$

$$= 121 \text{ g/mole}$$

Question 16

What is the density of SO_3 gas at 0°C and 0.980 atm?

Answer 16

3.50 g/L

$$pV = nRT = \frac{\#g}{GMW} \times RT$$

OEIOU: $\dfrac{\#g}{V} = D = \dfrac{p \times GMW}{RT}$

$$= \frac{(0.980 \text{ atm})(80.0 \text{ g/mole})}{(0.0821 \text{ L-atm/mole}°K)(273°K)}$$

$$= 3.50 \text{ g/L}$$

Question 17

A 0.500-g sample of an unknown organic liquid is vaporized at 80°C and 760 mm in a 125-ml flask. Could this liquid be benzene, C_6H_6?

Answer 17

No; the molecular weight is too high.

$$GMW_x = \frac{\#g \times RT}{pV}$$

$$= \frac{(0.500 \text{ g})(0.0821 \text{ L-atm/mole}°K)(353°K)}{(1.00 \text{ atm})(0.125 \text{ L})}$$

$$= 116 \text{ g/mole}$$

$$GMW_{C_6H_6} = 78$$

Thus $GMW_x \neq GMW_{C_6H_6}$

Statement 6

The ideal gas equation can also be used in solving **stoichiometric problems** involving gases, using the relationship

$$\#\text{moles} = n = pV/RT$$

Question 18

How many liters of CO_2 are produced at 25°C and 760 mm by burning 24.0 g of CH_4 in oxygen?

Answer 18

36.7 L

Reaction: $CH_4 + 2O_2 \rightarrow CO_2 + 2H_2O$

SS: $\#\text{moles } CO_2 = \#\text{moles } CH_4$

Substituting: $\dfrac{pV}{RT} = \dfrac{\#g\ CH_4}{GMW_{CH_4}}$

OEIOU: $V = \dfrac{\#g\ CH_4 \times RT}{GMW_{CH_4} \times p}$

$$= \frac{(24.0 \text{ g})(0.0821 \text{ L-atm/mole}°K)(298°K)}{(16.0 \text{ g/mole})(1.00 \text{ atm})}$$

$$= 36.7 \text{ L}$$

Question 19

How many liters of oxygen would be required to burn 24.0 g of methane at 25°C and 760 mm?

Answer 19

73.4 L

SS: # moles O_2 = 2(# moles CH_4)

OEIOU: $V = \dfrac{2 \times \text{\#g } CH_4 \times RT}{GMW_{CH_4} \times p}$

Question 20

When sufficiently heated, $Ca(OH)_2$ decomposes to give CaO and H_2O vapor. How many milliliters of water vapor would be produced by heating 0.761 g of calcium hydroxide at 600°C and 1.00 atm?

Answer 20

736 ml

Rx: $Ca(OH)_2 \rightarrow CaO + H_2O$

SS: # moles H_2O = # moles $Ca(OH)_2$

OEIOU: $V = \dfrac{\text{\#g } Ca(OH)_2 \times RT}{GMW_{Ca(OH)_2} \times p}$

= 0.736 L or 736 ml

Question 21

How many liters of hydrogen gas are required theoretically to produce 4.00 L of NH_3 on reaction of H_2 and N_2 at 25°C and 760 torr?

Answer 21

6.00 L

Rx: $N_2 + 3H_2 \rightarrow 2NH_3$

SS: # moles $H_2 = \dfrac{3}{2} \times$ # moles NH_3

Substituting: $\left(\dfrac{pV}{RT}\right)_{H_2} = \dfrac{3}{2} \times \left(\dfrac{pV}{RT}\right)_{NH_3}$

But p, R, and T are the *same* on both sides, so:

$$V_{H_2} = \dfrac{3}{2} \times V_{NH_3}$$

$$= 6.00 \text{ L}$$

(You may have realized this without doing the calculations.)

Statement 7

Dalton's Law states that the **total pressure** in a container is the sum of the **partial pressures** of the component gases present. Mathematically:

$$p_{total} = p_A + p_B + p_C + \ldots$$

where p_A, p_B, and p_C are the partial pressures.

A popular use of Dalton's Law is in correcting for the presence of water vapor in a gas collected "over water." A gas is produced by a chemical reaction and collected by bubbling it into an inverted bottle filled with water. The gas displaces the water but remains saturated with water vapor, which contributes its pressure to the total pressure inside the bottle.

Question 22

The total pressure of a mixture of N_2, H_2, and O_2 is 800 mm. If the partial pressures of N_2 and O_2 are 500 mm and 125 mm, respectively, what is the partial pressure of H_2?

Answer 22

175 mm

$$p_{total} = p_{N_2} + p_{H_2} + p_{O_2}$$

OEIOU: $p_{H_2} = p_{tot} - (p_{N_2} + p_{O_2})$

$= 800 \text{ mm} - (500 + 125) \text{ mm}$

$= 175 \text{ mm}$

Question 23

Hydrogen gas is produced by reacting zinc with H_2SO_4. The H_2 is collected by displacing water from a bottle. The gas has a volume of 250 ml, the pressure is 750 mm, and the temperature is 25°C. If the partial pressure (or "vapor pressure") of water is 24 mm at 25°C, how many moles of H_2 are formed?

Answer 23

0.00976 mole

$$P_{H_2}V = n_{H_2}RT$$

Substituting: $(p_{tot} - p_{H_2O}) = n_{H_2}RT$

OEIOU: $n_{H_2} = \dfrac{(p_{tot} - p_{H_2O})V}{RT}$

$$= \dfrac{\dfrac{750 \text{ mm} - 24 \text{ mm}}{760 \text{ mm/atm}}(0.250 \text{ L})}{(0.0821 \text{ L-atm/mole°K})(298°K)}$$

$= 0.00976$ mole

Statement 8

Graham's Law states that the rate of **effusion** of a gas is inversely proportional to the square root of its density. (Effusion is the leaking of a gas through a small hole into a vacuum.) For two different gases, A and B, under identical conditions:

$$\dfrac{(\text{rate})_A}{(\text{rate})_B} = \sqrt{\dfrac{D_B}{D_A}}$$

Or in terms of molecular weight:

$$\dfrac{(\text{rate})_A}{(\text{rate})_B} = \sqrt{\dfrac{GMW_B}{GMW_A}}$$

Question 24

Helium effuses through a small hole in a certain flask at the rate of 0.0200 mole/sec. Neon leaks through the same hole at 8.90×10^{-3} mole/sec. If the density of helium is 0.164 g/L, what is the density of neon?

Answer 24

0.828 g/L

$$\frac{(\text{rate})_{Ne}}{(\text{rate})_{He}} = \sqrt{\frac{D_{He}}{D_{Ne}}}$$

$$\frac{(\text{rate})_{Ne}^2}{(\text{rate})_{He}^2} = \frac{D_{He}}{D_{Ne}}$$

$$\text{OEIOU:} \quad D_{Ne} = D_{He} \times \frac{(\text{rate})_{He}^2}{(\text{rate})_{Ne}^2}$$

Question 25

An unknown organic gas effuses at a rate of 0.0100 mole/sec. Methane (CH_4) has a rate of effusion of 0.0137 mole/sec under the same conditions. What is the molecular weight of the unknown gas?

Answer 25

30.0 g/mole

$$\frac{(\text{rate})_x}{(\text{rate})_{CH_4}} = \sqrt{\frac{(GMW)_{CH_4}}{(GMW)_x}}$$

$$\text{OEIOU:} \quad (GMW)_x = (GMW)_{CH_4} \times \frac{(\text{rate})_{CH_4}^2}{(\text{rate})_x^2}$$

Question 26

An unknown gas requires 50 sec to leak through a small hole into a vacuum. The same number of moles of oxygen requires 63 sec. What is the molecular weight of the gas?

Answer 26

20 g/mole

$$\frac{(\text{rate})_x}{(\text{rate})_{O_2}} = \sqrt{\frac{GMW_{O_2}}{GMW_x}}$$

But: $\dfrac{(\text{rate})_x}{(\text{rate})_{O_2}} = \dfrac{(\text{time})_{O_2}}{(\text{time})_x}$

Thus: $\dfrac{(\text{time})_{O_2}}{(\text{time})_x} = \sqrt{\dfrac{GMW_{O_2}}{GMW_x}}$

Note that rate and time are *inversely* related; the faster the rate, the less time required.

OEIOU: $GMW_x = GMW_{O_2} \times \dfrac{(\text{time})_x^2}{(\text{time})_{O_2}^2}$

Statement 9

The ideal gas equation is exact only for an **ideal gas**—one in which the molecules or atoms composing it take up no space and exert no influence on one another.

In a **real gas** the molecules or atoms do take up space and influence one another; so in order to arrive at more precise agreement between predictions and actual measurements of real gases, more complex equations have been developed. One of these is the **van der Waals equation**.

This equation assumes that for one mole of a real gas $p_{ideal} = p + \dfrac{a}{V^2}$ and $V_{ideal} = V - b$, so that:

$$\left(p + \dfrac{a}{V^2}\right)(V - b) = RT$$

where a and b are constants dependent upon attractive forces between molecules or atoms of the gas (a) and the volume taken up by them (b); p is the measured pressure; V is the actual volume; and p_{ideal} and V_{ideal} are the pressure and volume the gas would have if it behaved as an ideal gas.

Question 27

Given that for helium:

a = 3.54 × 10^{-2} atm (L/mole)^2

b = 2.36 × 10^{-2} L/mole

calculate p_{ideal} and V_{ideal} for one mole of He gas at 0°C and 1.000 atm if the actual volume is 22.47 L.

Answer 27

p_{ideal} = 1.000 atm

V_{ideal} = 22.45 L

$$p_{ideal} = p + \frac{a}{V^2}$$

$$= 1.000 \text{ atm} + \frac{3.54 \times 10^{-2} \text{ atm (L/mole)}^2}{(22.47 \text{ L})^2}$$

$$= 1.000 \text{ atm} + 0.00007 \text{ atm}$$

$$\approx 1.000 \text{ atm}$$

$$V_{ideal} = V - b$$

$$= 22.47 \text{ L} - 0.024 \text{ L}$$

$$= 22.45 \text{ L}$$

Question 28

For one mole of chlorine:

a = 6.65 atm (L/mole)^2

b = 5.60 × 10^{-2} L/mole

At 25°C and 1.000 atm pressure the actual volume occupied by a mole of Cl_2 is 24.08 liters. Compare the calculated value of (RT) using:

a. the ideal gas equation

b. the van der Waals equation.

Answer 28

a. (RT) = 24.08 L-atm

OEIOU: (RT) = pV

$$= (1.000 \text{ atm})(24.08 \text{ L})$$

$$= 24.08 \text{ L}$$

b. (RT) = 24.30 L-atm

OEIOU: (RT) = $(p + \frac{a}{V^2})(V - b)$

$$= \left[1.000 \text{ atm} + \frac{6.65 \text{ atm (L/mole)}^2}{(24.08)^2}\right](24.08 \text{ L} - 0.56 \text{ L/mole})$$

$$= 24.30 \text{ L-atm}$$

This value of RT ideally should be independent of the gas, the volume, and the pressure. We see here that the van der Waals value (24.30 L) comes closer than the ideal gas value (24.08 L) to the value of 24.4 L calculated by multiplying R and T:

$$RT = (0.0821 \text{ L-atm/mole }°K)(298°K)$$
$$= 24.4 \text{ L}$$

Statement 10

At any given instant there is a broad distribution of velocities for the molecules or atoms of a gas, ranging from a few slow-moving ones through a large number with intermediate velocities to a few fast-moving ones.

The **average velocity** (also called "root mean square velocity") of the molecules or atoms of a gas can be calculated from the equation

$$u = \sqrt{\frac{3RT}{(GMW)}}$$

where u is the average velocity and R is expressed in ergs/mole°K—$R = 8.31 \times 10^7$ erg/mole°K. (1 erg = 1 gm cm²/sec².)

Question 29

Air is approximately 78% nitrogen by volume. What is the velocity of an average N_2 molecule in the air at "room temperature" (25°C)?

Answer 29

5.15×10^4 cm/sec

OEIOU: $u = \sqrt{\dfrac{3RT}{GMW_{N_2}}}$

$= \sqrt{\dfrac{(3)(8.31 \times 10^7 \text{ erg/mole}°K)(298°K)}{28.01 \text{ g/mole}}}$

$= 5.15 \times 10^4$ cm/sec

Question 30

What is the average velocity of a $CHCl_3$ molecule in chloroform vapor at 75°C?

Answer 30

2.69×10^4 cm/sec

TESTING YOUR MASTERY—Part 10   Gases

Question 31

A certain gas is confined at 33°C and 778 mm Hg. Express these conditions in °K and atmospheres.

Answer 31

306°K and 1.02 atm

OEIOU: °K = °C + 273°

OEIOU: # atm = # mm Hg $\times \dfrac{\text{atm}}{\text{mm Hg}}$

Question 32

A sample of nitrogen gas occupies 214 ml at 27°C and 760 torr. If the pressure is changed to 840 torr, what will be the new volume occupied by the nitrogen?

Answer 32

194 ml

OEIOU: $V_2 = V_1 \times \dfrac{p_1}{p_2}$

(Units of p are not significant.)

Question 33

A sample of nitrogen gas occupies 214 ml at 27°C and 760 torr. If the temperature is changed to -1°C, what will be the new volume occupied by the nitrogen?

Answer 33

194 ml

OEIOU: $V_2 = V_1 \times \dfrac{T_2}{T_1}$

(Temperature units must be °K.)

Question 34

A sample of nitrogen gas confined at 27°C and 1.00 atm occupies 214 ml. The gas is cooled to −1°C and the pressure increased to 840 mm Hg. What will happen to the volume of the gas?

Answer 34

The volume will shrink to 176 ml.

OEIOU: $V_2 = V_1 \times \dfrac{T_2}{T_1} \times \dfrac{p_1}{p_2}$

(The same units must be used for p_1 and p_2).

Question 35

A mole of an ideal gas occupies 22.4 L at 0°C and 1.00 atm pressure. If the temperature is held constant, to what must the pressure be changed to give a volume of 10.0 liters?

Answer 35

2.24 atm (or 1702 torr)

OEIOU: $p_2 = p_1 \times \dfrac{V_1}{V_2}$

Question 36

Suppose the pressure had been held constant in Question 35, but the temperature was changed. What new temperature would be required to cause the volume of the mole of gas to shrink to 10.0 liters?

Answer 36

122° K (or −151 °C)

OEIOU: $T_2 = T_1 \times \dfrac{V_2}{V_1}$

Question 37

A sample of oxygen gas occupies 308 ml at 35°C and 1.50 atm. The temperature is changed to 45°C. To what must the pressure be changed to increase the volume by 50 ml?

Answer 37

1.33 atm

$$\text{OEIOU:} \quad p_2 = \frac{V_1}{(V_1 + \Delta V)} \times \frac{T_2}{T_1} \times p_1$$

Question 38

If the temperature of a gas sample is changed from 0°C to −20°C, and if the original pressure was 0.940 atm, to what must the pressure be changed to prevent the volume from changing?

Answer 38

0.871 atm

$$\text{OEIOU:} \quad p_2 = \frac{T_2}{T_1} \times p_1$$

Question 39

A certain sample of neon gas at 25°C and 1.33 atm occupies 10.0 liters. How many moles of neon are present?

Answer 39

0.543 mole

$$\text{OEIOU:} \quad n = PV/RT$$

Question 40

Assuming that all the gases involved behave ideally, what would be the approximate volume change when the following reaction takes place at 1.00 atm and 25°C?

$$2N_2O(g) \rightarrow 2N_2(g) + O_2(g)$$

PART 10—GASES

Answer 40

24.4 L

OEIOU: $\Delta V = RT/P$

Question 41

At what pressure would 0.346 mole of helium occupy 10.0 liters at 0°C?

Answer 41

0.776 atm

OEIOU: $P = \dfrac{nRT}{V}$

Question 42

At what temperature would 0.346 mole of helium at 1.00 atm occupy 10.0 liters?

Answer 42

352°K (or 79°C)

OEIOU: $T = \dfrac{pV}{nR}$

Question 43

A 0.500-g sample of octane is collected as 144 ml of a gas at 126°C and 760 torr. What is the molecular weight of octane?

Answer 43

114 g/mole

OEIOU: $GMW = \#g \times \dfrac{RT}{pV}$

Question 44

If NO_2 behaves ideally, what should be its density at 25°C and 1.00 atm?

Answer 44

1.88 g/L

OEIOU: $D = \dfrac{p \times GMW}{RT}$

Question 45

How many liters of O_2 gas are produced at 25°C and 1.00 atm when O_2 is generated by decomposing 10.0 g of $KClO_3$:

$$2KClO_3(s) \rightarrow 2KCl(s) + 3O_2(g)$$

Answer 45

2.99 L

OEIOU: $V = \dfrac{3}{2} \times \dfrac{\#g\ KClO_3 \times RT}{GMW_{KClO_3} \times p}$

Question 46

How many liters of H_2 gas at 100°C and 1.00 atm are required to reduce 10.0 g of CuO to Cu metal according to the following equation:

$$CuO(s) + H_2(g) \rightarrow H_2O(g) + Cu(s)$$

Answer 46

3.85 L

OEIOU: $V = \dfrac{\#g\ CuO \times RT}{GMW_{CuO} \times p}$

Question 47

How many liters of N_2 are required to react with O_2 at 150°C and 780 mm Hg in order to produce 10.0 liters of NO?

Answer 47

5.00 L

OEIOU: $V_{N_2} = \dfrac{1}{2} \times V_{NO}$

Question 48

A mixture of CO, CO_2, and O_2 is confined in a 275-ml flask at 0°C. If the total pressure is 780 torr, and if the CO and CO_2 each exert 330 torr, what is the pressure exerted by the O_2?

Answer 48

120 torr

OEIOU: $P_{O_2} = P_{tot} - (P_{CO} + P_{CO_2})$

Question 49

The O_2 gas produced by decomposing $KClO_3$ is collected by displacing water from a bottle. A total of 312 ml of water is displaced at 24°C and 747 torr. If the vapor pressure of water is 22.4 torr at 24°C,

a. how many moles of O_2 were produced?

b. how many grams of O_2 were produced?

Answer 49

a. 0.0124 mole

OEIOU: $n_{O_2} = \dfrac{(P_{tot} - P_{H_2O}) V}{RT}$

b. 0.397 g

OEIOU: $\#g_{O_2} = \dfrac{(P_{tot} - P_{H_2O}) V}{RT} \times GMW_{O_2}$

Question 50

An unknown organic gas has an effusion rate of 0.0121 mole/sec as compared to a rate of 0.0153 mole/sec for N_2 in the same apparatus. What is the molecular weight of the organic compound?

Answer 50

44.8 g/mole

OEIOU: $GMW_x = GMW_{N_2} \times \dfrac{(rate)^2_{N_2}}{(rate)^2_x}$

Question 51

An unknown gas requires 93 sec to effuse in a certain apparatus. The same number of moles of N_2 require 123 sec under the same conditions. What is the GMW of the compound?

Answer 51

16.0 g/mole

$$\text{OEIOU: } GMW_x = GMW_{N_2} \times \frac{(\text{time})_x^2}{(\text{time})_{N_2}^2}$$

Question 52

The van der Waals factors for one mole of CO_2 are:

a = 3.68 atm $(L/mole)^2$

b = 4.25 × 10^{-2} L/mole

If the actual volume of a mole of CO_2 is 22.29 L at $0°C$ and 1.00 atm, compare the value of RT as calculated by:

a. the ideal gas equation.

b. the van der Waals equation.

Answer 52

a. 22.29 L-atm

$$\text{OEIOU: } RT = pV$$

b. 22.41 L-atm

$$\text{OEIOU: } RT = (p + \frac{a}{V_2^2})(V - b)$$

Question 53

What is the average velocity of a molecule of CO_2 at $0°C$?

Answer 53

1.24 × 10^4 cm/sec

$$\text{OEIOU: } u = \sqrt{\frac{3\,RT}{GMW_{CO_2}}}$$

(Remember to express R in "erg/mole $°K$.")

Question 54

To what temperature would CO_2 have to be heated to give a root mean square velocity of 4.00×10^4 cm/sec?

Answer 54

282°K (or 9°C)

OEIOU: $T = \dfrac{u^2 \times GMW_{CO_2}}{3R}$

PART 11
LIQUIDS

Statement 1

In the **liquid state** the molecules (or atoms) of a substance are very close together, but they slide past one another with relative ease. If energy is supplied, the liquid can be vaporized (i.e., converted into a gas).

The amount of heat necessary to convert **one mole** of a liquid into a vapor at a given temperature (usually its normal boiling point; see Statement 2 in this section) is called the **heat of vaporization**, ΔH_{vap}. The value of ΔH_{vap} increases with the strength of the intermolecular forces that must be overcome to convert the liquid into a vapor. The following **general rules** are helpful in predicting the relative strengths of intermolecular forces.

a. Intermolecular forces include the following:

   H-bonds—between adjacent polar molecules of the type $\overset{\delta+\delta-}{\text{H-Y}}$, where Y is very electronegative (principally F, O, and N for "true" H-bonds). H will be more positive than Y, and will be attracted to the more negative Y of adjacent molecule.

   dipole-dipole bonds—between polar molecules. The more negative end of one molecule will be attracted to the more positive end of another.

   van der Waals forces—between adjacent nonpolar molecules. Result from temporary dipoles in the molecules. Present in all kinds of substances, but most important for nonpolar substances.

b. For H-bonded molecules, the strength of intermolecular forces increases with the electronegativity of Y.

c. For other polar molecules, the strength of intermolecular forces increases with increasing polarity.

d. For nonpolar molecules, the strength of intermolecular forces increases with molecular weight.

Question 1

What kinds of forces must be overcome in order to vaporize these liquids?

a. CCl_4

b. H_2O

c. $CHCl_3$

Answer 1

a. van der Waals forces

b. H-bonds

c. dipole-dipole bonds

Van der Waals forces are operating in (b) and (c) also, but most of the energy used to vaporize H_2O and $CHCl_3$ is used to overcome H-bonds and dipole-dipole forces.

Question 2

Which of these liquids has the higher heat of vaporization?

a. liquid CH_4

b. liquid SiH_4

Answer 2

b. liquid SiH_4

The compounds are tetrahedral (see Part 4), thus non-polar, and the molecular weight of SiH_4 is greater than that of CH_4.

Question 3

Which of these liquids has the higher ΔH_{vap}?

a. liquid H_2O

b. liquid H_2S

Answer 3

a. liquid H_2O

H_2O is strongly H-bonded; H_2S is polar but not H-bonded to any significant extent (see rule *a* in Statement 1 of this section).

Note: Although one end of the H–S dipole is a hydrogen, the resulting attraction is too weak to be considered an H-bond.

Statement 2

Upon reaching its **boiling point**, a liquid will boil (vaporize) continuously until all the liquid has vaporized, with the temperature holding constant at the boiling point. When the confining pressure is 1 atm, we speak of the "normal boiling point," T_b.

Trouton's Rule relates ΔH_{vap} and T_b:

$$\text{entropy of vaporization} = S_{vap} = \frac{\Delta H_{vap}}{T_b} \approx 21 \text{ cal/mole}°K$$

Question 4

The normal boiling point of ethyl alcohol is 78.5°C. Predict its heat of vaporization.

Answer 4

7.4×10^3 cal/mole

$$\Delta H_{vap}/T_b \approx 21 \text{ cal/mole}°K$$

OEIOU: $\Delta H_{vap} \approx T_b \times 21$ cal/mole°K

$= 351.5°K \times 21$ cal/mole°K

$= 7.4 \times 10^3$ cal/mole

(Compare this predicted value with the actual value of 9.22×10^3 cal/mole.)

Question 5

If T_b for chloroform is 61.3°C, predict its heat of vaporization.

Answer 5

7.0×10^3 cal/mole

(Experimental value is 7.02×10^3 cal/mole.)

Question 6

The heat of vaporization for benzene is 7.35×10^3 cal/mole °K. Predict its normal boiling point.

Answer 6

77°C (or 350°K)

OEIOU: $T_b \approx \dfrac{\Delta H_{vap}}{21 \text{ cal/mole}°K}$

(Experimental value is 80.1 °C.)

Question 7

Arrange these liquids in order of decreasing boiling point:

NH_3 PH_3 AsH_3

Answer 7

$NH_3 > AsH_3 > PH_3$

Same considerations as for heat of vaporization. Only NH_3 undergoes significant H-bonding. AsH_3 and PH_3 are both trigonal pyramidal molecules, with AsH_3 having the greater molecular weight.

Statement 3

At a given temperature an **equilibrium** is established between molecules leaving the surface of a liquid and those returning from the gaseous state above it. At equilibrium, shown as:

$$\text{substance (l)} \rightleftharpoons \text{substance (g)}$$

molecules are leaving and returning at the same rate, so that there is no net change. The vapor molecules above the liquid exert a pressure called the **vapor pressure**, p_{vap}, which is characteristic of the substance, and which increases as temperature increases. The boiling point is defined as the temperature at which p_{vap} and the confining pressure (usually atmospheric) are equal.

The change in p_{vap} with T can be predicted from the **Clausius-Clapeyron equation**:

$$\log \frac{p_2}{p_1} = \frac{\Delta H_{vap}}{2.303 R} \left(\frac{T_2 - T_1}{T_2 T_1} \right)$$

where p_1 and p_2 are the vapor pressures corresponding to the temperatures T_1 and T_2. In this equation, R is expressed in cal/mole °K: $R = 1.99$ cal/mole °K.

Question 8

The vapor pressure of water at 25°C is 23.8 mm. What will be its p_{vap} at 60°C if its heat of vaporization is 10.4 kcal/mole?

Answer 8

150 mm

$$\log \frac{p_2}{p_1} = \frac{\Delta H_{vap}}{2.303 R} \left(\frac{T_2 - T_1}{T_2 T_1} \right)$$

$$\text{OEIOU: } p_2 = p_1 \times \text{antilog} \left[\frac{\Delta H_{vap}}{2.303\, R} \left(\frac{T_2 - T_1}{T_2 T_1} \right) \right]$$

$$= p_1 \times \text{antilog} \left[\frac{10.4 \times 10^3 \text{ cal/mole } (333-298)°K}{2.303 \times 1.99 \text{ cal/mole}°K \times 333° \times 298°K} \right]$$

$$= 150 \text{ mm}$$

Question 9

If ΔH_{vap} for CCl_4 is 7.17 kcal/mole and its p_{vap} is 305 torr at 50°C, what will be its p_{vap} at "room temperature" (25°C)?

Answer 9

119 torr

Statement 4

By increasing the pressure on a liquid, its temperature can be raised above T_b, its normal boiling point. There is a maximum temperature, called the **critical temperature**, T_c, above which a substance will not exist as a liquid but will vaporize regardless of the applied pressure. Its relation to T_b is as follows:

$$T_c \approx \frac{3}{2} T_b$$

Question 10

The normal boiling point of ether is 34.6°C. Predict its critical temperature.

Answer 10

188°C (or 461°K)

$$\text{OEIOU: } T_c \approx \frac{3}{2} T_b$$

$$= \frac{3}{2} \times 307.6°K$$

$$= 461°K \text{ (or } 188°C)$$

Question 11

Predict T_c for mercury, knowing that its normal boiling point is 357°C.

Answer 11

672°C

TESTING YOUR MASTERY—Part 11   Liquids

Question 12

What kinds of forces must be overcome to convert these liquids into vapor form?

a. liquid HF

b. liquid SiF_4

c. liquid CHF_3

Answer 12

a. H-bonds

b. van der Waals forces

c. dipole-dipole bonds

Question 13

In each of the following pairs of compounds, predict which should have the higher ΔH_{vap}:

a. liquid F_2 or liquid Br_2

b. liquid HF or liquid HBr

Answer 13

a. Br_2 (stronger van der Waals forces)

b. HF (H-bonds versus dipole-dipole bonds)

PART 11—LIQUIDS

Question 14

Arrange these liquids in order of decreasing boiling point expected:

$$GeCl_4 \quad CCl_4 \quad SiCl_4$$

Answer 14

$$GeCl_4 > SiCl_4 > CCl_4$$

All are nonpolar. The preceding is the order of decreasing strength of van der Waals forces.

Question 15

If the normal boiling point of octane is 126°C, what should its ΔH_{vap} be?

Answer 15

8.4×10^3 cal/mole

OEIOU: $\Delta H_{vap} \approx T_b \times 21$ cal/mole°K

Question 16

The normal boiling point of octane is 126°C. Predict its critical temperature.

Answer 16

326°C

OEIOU: $T_c \approx \frac{3}{2} T_b$

Question 17

The heat of vaporization of carbon tetrachloride is 7.17×10^3 cal/mole. Predict its normal boiling point.

Answer 17

341°K (or 68° C)

OEIOU: $T_b \approx \dfrac{\Delta H_{vap}}{21 \text{ cal/mole}^\circ K}$

(Experimental value is 350°K.)

Question 18.

The vapor pressure of ether at 20°C is 0.582 atm. Its heat of vaporization is 6.2×10^3 cal/mole. What is the vapor pressure of ether at 0°C?

Answer 18

0.267 atm

OEIOU: $p_2 = p_1 \times \text{antilog} \left[\dfrac{\Delta H_{vap}}{2.303\, R} \left(\dfrac{T_2 - T_1}{T_2 T_1} \right) \right]$

PART 12
SOLIDS

Statement 1

In the **solid state** a substance assumes an orderly **crystalline** structure. The ions, molecules, or atoms composing the crystal are very restricted in their freedom to move. When enough energy is supplied, however, a solid can be converted into a liquid (i.e., melted).

The temperature at which solid and liquid are in equilibrium is called the **melting point**, T_m. The amount of heat required to melt one mole of a solid at a constant temperature (usually its normal melting point) is called the molar **heat of fusion** of the substance.

The melting point for molecular substances, therefore, varies according to whether the forces holding them together are H-bonds (strongest), dipole-dipole bonds, or van der Waals forces (weakest).

Question 1

Which of these compounds has the higher melting point:

Br_2 or I_2?

Answer 1

I_2

I_2 has the higher molecular weight; thus stronger van der Waals forces.

Question 2

Which of these compounds should have the higher melting point:

N_2 or CO?

Answer 2

CO

They have the same molecular weight, but CO is polar (dipole-dipole bonds), and N_2 is nonpolar (van der Waals forces).

Question 3

Predict whether H_2O or H_2S should have the higher T_m.

Answer 3

H_2O

Strong H-bonding in H_2O only.

Statement 2

The melting points of **ionic compounds** increase with

a. increasing charge on the ions;

b. decreasing size of the ions (sum of ionic radii).

The melting points of ionic compounds are higher than those of molecular compounds because of the strong forces of attraction between oppositely-charged ions.

Question 4

Which of these compounds should have the higher melting point:

NaCl or KCl ?

Answer 4

NaCl

Same charges on ions, but we know that Na^+ is smaller than K^+, because it lies above K in Group 1A.

Question 5

The radius of Li^+ is 0.60 Å; the radius of Ca^{2+} is 0.99 Å. The Cl^- ion has a radius of 1.81 Å, and the radius of O^{2-} is 1.40 Å. Predict whether LiCl or CaO will have the higher melting point.

Answer 5

CaO

$r_{Li^+} + r_{Cl^-} = 2.41$ Å

$r_{Ca^{2+}} + r_{O^{2-}} = 2.39$ Å

There is not much difference in the sum of the ionic radii; LiCl slightly larger. But the *charges* (or ions) in CaO are +2 and −2; in LiCl ions they are only +1 and −1.

Question 6

Arrange these compounds in order of decreasing T_m:

$$CaCl_2 \quad MgCl_2 \quad CCl_4$$

Answer 6

$$MgCl_2 > CaCl_2 > CCl_4$$

The same charges on ions are present in $MgCl_2$ and $CaCl_2$, but Mg^{2+} is *smaller* than Ca^{2+}. $CaCl_2$ is *ionic*; CCl_4 is covalent (molecular).

Question 7

Arrange these compounds in order of decreasing T_m:

$$CO_2 \quad SrS \quad CS_2$$

Answer 7

$$SrS > CS_2 > CO_2$$

Statement 3

Solids, like liquids, have a vapor pressure that results from molecules (or atoms) escaping from the solid and passing directly into the gaseous phase. The conversion of a substance from a solid into a vapor is called **sublimation**.

The relationships among solid, liquid, and vapor phases of a pure substance are shown conveniently by means of a **phase diagram**. This is basically a graphical plot of vapor pressure versus temperature for solid and liquid phases, with a plot of p versus T_m shown on the same graph.

The phase diagram is thus divided by lines into regions where the substance can exist as:

solid only — p too high for liquid or vapor to exist at the specified T.

liquid only — p too high for vapor to exist, too low for solid to exist at the specified T.

vapor only — p too low for liquid or solid to exist at the specified T.

solid and vapor in equilibrium — at p and T combinations on the line separating **solid only** and **vapor only** regions.

222   PART 12—SOLIDS

solid and liquid in equilibrium — at p and T combinations on the line separating **solid only** and **liquid only** regions.

solid, liquid, and vapor in equilibrium — at the one point where the three lines separating solid, liquid, and vapor regions intersect; called the **triple point** of the substance.

Questions 8–12 are based on this phase diagram (not to scale) for water.

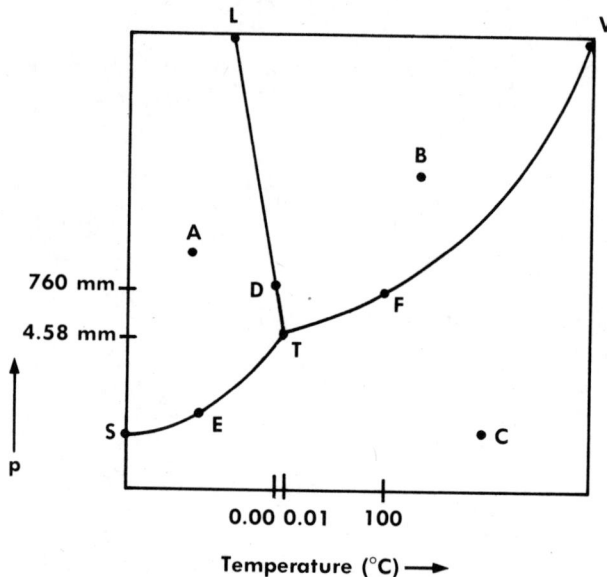

Question 8

At what labeled point does water exist as:

a. ice (solid) only?

b. liquid only?

c. steam (vapor) only?

> **Answer 8**
>
> a. point A (in region STL)
>
> b. point B (in region LTV)
>
> c. point C (in region STV)

Question 9

At what labeled point are the following in equilibrium:

a. liquid and vapor

b. solid and liquid

c. solid and vapor

Answer 9

a. point F (on line TV)

b. point D (on line LT)

c. point E (on line ST)

Question 10

What point is:

a. the normal melting point of water?

b. the triple point of water?

Answer 10

a. point D (0.00° C, 760 mm)

T_m is by definition the melting point at 1 atm (760 mm).

b. point T (0.01° C, 4.58 mm)

This is the only point at which solid, liquid, and vapor are in equilibrium.

Question 11

What will happen if:

a. the temperature is increased from 0.00°C to 50°C at p = 760 mm?

b. the pressure is changed from 760 mm to 4.58 mm at temperature = 50°C?

Answer 11

a. All the solid melts; that is, it shifts into region LTV, which is liquid only.

b. All the liquid vaporizes; that is, it shifts into region STV, which is vapor only.

Question 12

Will the melting point increase or decrease with increasing pressure?

Answer 12

Decrease.

The LT line shows what happens to the melting point as p changes. In the case of water, as p increases along this line, the melting point drops in value. (For many substances, however, the melting point increases as p increases.)

TESTING YOUR MASTERY—Part 12   Solids

Question 13

Arrange these hydrides in order of decreasing melting points:

CH_4 NH_3 H_2O

Answer 13

$H_2O > NH_3 > CH_4$

H-bonds are stronger in H_2O than in NH_3. Van der Waals forces only are in CH_4.

Question 14

Arrange these iodides in order of decreasing melting points:

CI_4 LiI   CsI

Answer 14

LiI $>$ CsI $>$ CI_4

LiI and CsI are ionic, with r_{Li^+} smaller than r_{Cs^+}. CI_4 is held together by weaker van der Waals forces.

Question 15

Arrange these calcium salts in order of decreasing melting point:

CaO   CaS   CaSe

Answer 15

CaO > CaS > CaSe

All are ionic, with $r_{O^{2-}} < r_{S^{2-}} < r_{Se^{2-}}$.

Question 16

The radii of Na^+ and Ca^{2+} are 0.95 Å and 0.99 Å, respectively. The radii of Br^- and Se^{2-} are 1.95 Å and 1.98 Å, respectively. Would you expect NaBr, or CaSe, to have the higher melting point?

Answer 16

CaSe

The sum of the ionic radii is not very different, but the charges on Ca^{2+} and Se^{2-} ions are larger than those on Na^+ and Br^-.

Questions 17–22 refer to this phase diagram (not to scale) for carbon dioxide:

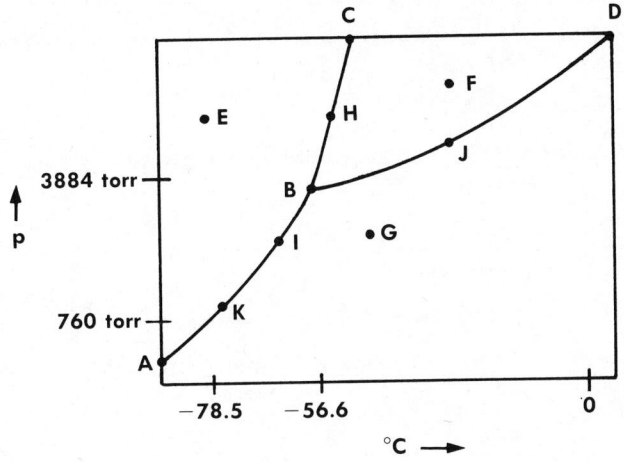

Question 17

At what labeled point does CO_2 exist only as a:

a. vapor?

b. liquid?

c. solid?

Answer 17

a. G

b. F

c. E

Question 18

At what labeled points are the following phases in equilibrium:

a. vapor and solid

b. liquid and vapor

c. solid and liquid

d. solid, liquid, and vapor

Answer 18

a. I

b. J

c. H

d. B

Question 19

Will the melting point of CO_2 increase or decrease as pressure increases?

Answer 19

Increase.

The line BC slants to the right, toward higher temperatures.

Question 20

What will happen when:

a. the pressure of CO_2 is increased from 760 torr to 3884 torr at $-77°C$?

b. the temperature of CO_2 is increased from $-78.5°C$ to $-56.6°C$ at 780 torr?

c. the temperature of CO_2 is increased from $-78.5°C$ to $-50°C$ at 3890 torr?

Answer 20

a. The CO_2 changes from vapor to solid.

b. The CO_2 changes from solid to vapor (i.e., it "sublimes").

c. The CO_2 changes from solid to liquid (i.e., it melts).

Question 21

What is the triple point of CO_2?

Answer 21

point B $(-56.6°C)$

Question 22

In what state does CO_2 exist at $25°C$ and 1.00 atm?

Answer 22

vapor

PART 13
SOLUTIONS

Statement 1

A **solution** is a homogeneous mixture of two or more substances.

By convention, if one component of a solution is a gas or solid and the other is a liquid, the liquid is called the **solvent** and the gas or solid the **solute**. When both components are liquid, the one present in greater quantity is called the solvent.

Question 1

For each of these mixtures, which component is considered to be the solvent?

a. 25 g NaCl and 100 g H_2O

b. 100 g benzene and 0.5 g aspirin

Answer 1
a. H_2O
b. benzene

Question 2

Which component of these mixtures is considered to be the solute?

a. 2.5 g CO_2 and 1000 g H_2O

b. 10 g NaCl, 200 g H_2O, and 3 g HCl

Answer 2
a. CO_2
b. both NaCl and HCl

Question 3

When 100 g of H_2O and 500 g of ethyl alcohol are mixed, which component is considered to be the solvent?

Answer 3

ethyl alcohol

Statement 2

The **concentration** of a solute in a solution may be expressed in several ways.

The **weight per cent** of a solute is given by:

$$\%A = \frac{\text{wt } A}{\text{wt } A + \text{wt } B + \ldots} \times 100$$

where wt A, wt B, and so forth, are the weights of each component (in two-component solutions this will be merely the weights of solute and solvent).

Question 4

If 11.9 g of KBr is dissolved in 500.0 g of water, what is the weight per cent of KBr in the solution?

Answer 4

2.32 % KBr

$$\text{OEIOU: } \%\text{KBr} = \frac{\text{g KBr}}{\text{g KBr} + \text{g H}_2\text{O}} \times 100$$

Question 5

5.00 g of aspirin ($C_9H_8O_4$) is dissolved in 100 ml (density = 0.789 g/ml) of ethyl alcohol (C_2H_5OH). What is the weight per cent of aspirin in the solution?

Answer 5

5.96 %

$$\text{OEIOU: } \%\text{ Asp} = \frac{\text{g Asp}}{\text{g Asp} + (\text{ml}_{et} \times D_{et})} \times 100$$

Statement 3

The **per cent by volume** of a solute is given by:

$$\text{vol.}\%_A = \frac{V_A}{V_A + V_B + \ldots} \times 100$$

where V_A, V_B, and so forth, are the volumes of the components mixed to form the solution.

Question 6

If 100.0 ml of water and 150.0 ml of ethyl alcohol are mixed, what will be the per cent by volume of ethyl alcohol in the solution?

Answer 6

60.0%

$$\text{Vol.}\%_{et} = \frac{V_{et}}{V_{et} + V_{H_2O}} \times 100$$

$$= \frac{150 \text{ ml}}{150 \text{ ml} + 100 \text{ ml}} \times 100$$

$$= 60.0\%$$

Question 7

If 100.0 g of water (D = 1.00 g/ml) and 150.0 g of ethyl alcohol (D = 0.789 g/ml) are mixed, what will be the per cent by volume of ethyl alcohol in the solution?

Answer 7

65.5%

$$\text{OEIOU: Vol.}\%_{et} = \frac{(\text{gm/D})_{et}}{(\text{gm/D})_{et} + (\text{gm/D})_{H_2O}} \times 100$$

Statement 4

To find the concentration of a solute in terms of its **mole fraction** (that is, the fraction of the total moles of material present that comprises the solute), we may use the expression:

PART 13—SOLUTIONS

$$X_A = \frac{n_A}{n_A + n_B + \ldots}$$

where n_A and n_B represent the number of moles of A and B, respectively. The total of the mole fractions of all constituents is **one**:

$$X_A + X_B + \ldots = 1.00 \ldots$$

Question 8

1.00 mole of ethyl alcohol is dissolved in 9.00 moles of water. What is the mole fraction of ethyl alcohol in the solution?

Answer 8

0.100

OEIOU: $X_{et} = \dfrac{n_{et}}{n_{et} + n_{H_2O}}$

Question 9

If 11.9 g of KBr is dissolved in 500.0 g of water, what are the mole fractions of KBr and of H_2O in the solution?

Answer 9

0.00359 KBr   0.99641 H_2O

OEIOU: $X_{KBr} = \dfrac{(g/GMW)_{KBr}}{(g/GMW)_{KBr} + (g/GMW)_{H_2O}}$

OEIOU: $X_{H_2O} = 1.00 \ldots - X_{KBr}$

Question 10

How much water should be added to 11.68 g of NaCl to give a solution in which the mole fraction of NaCl is 0.0200?

Answer 10

176 g H_2O

Defining equation: $X_{NaCl} = \dfrac{n_{NaCl}}{n_{NaCl} + n_{H_2O}}$

Substituting: $X_{NaCl} = \dfrac{(g/GMW)_{NaCl}}{(g/GMW)_{NaCl} + (g/GMW)_{H_2O}}$

$$X_{NaCl}(g/GMW)_{NaCl} + X_{NaCl}(g/GMW)_{H_2O} = (g/GMW)_{NaCl}$$

$$X_{NaCl}(g/GMW)_{H_2O} = (g/GMW)_{NaCl}(1 - X_{NaCl})$$

OEIOU: $g_{H_2O} = (g/GMW)_{NaCl}(1 - X_{NaCl}) \times \dfrac{GMW_{H_2O}}{X_{NaCl}}$

Question 11

How many grams of NaCl should be added to 100.0 g of water to give a solution in which the mole fraction of NaCl is 0.0200?

Answer 11

6.62 g NaCl

OEIOU: $g_{NaCl} = \dfrac{X_{NaCl} \times g_{H_2O} \times GMW_{NaCl}}{(1 - X_{NaCl}) \times GMW_{H_2O}}$

Statement 5

The most widely used measure of concentration is **molarity (M)**, the number of moles of solute per liter of solution, given by:

$$M_A = \dfrac{n_A}{\text{liters}_{solution}}$$

The volumes of solid and gaseous solutes are usually negligible in calculations involving liters of solution. The volume of a liquid solute must be taken into account; that is, when liquids $A, B, \ldots$ are mixed:

$$\text{liters}_{solution} = L_A + L_B + \ldots$$

Question 12

If 0.75 mole of HCl is dissolved in 1.50 liters of water, what is the molarity of the hydrochloric acid solution formed?

Answer 12

0.50 M

OEIOU: $M_{HCl} = n_{HCl}/L_{solution}$

$\qquad = (0.75 \text{ mole})/(1.50 \text{ L})$

$\qquad = 0.50 \text{ mole}/L$

or 0.50 M

Question 13

If 11.9 g of KBr is dissolved in 0.500 L of water, what is the molarity of the solution?

Answer 13

0.200 M

OEIOU: $M_{KBr} = \dfrac{g_{KBr}/GMW_{KBr}}{L_{solution}}$

Question 14

If 8.00 g of sodium hydroxide is dissolved in 250.0 ml of water, what is the molarity of the resulting NaOH solution?

Answer 14

0.800 M

Question 15

How many grams of NaCl should be added to one liter of water to produce a 0.125 M solution of NaCl?

Answer 15

7.31 g NaCl

Defining equation: $M_{NaCl} = \dfrac{n_{NaCl}}{L_{solution}}$

$\qquad\qquad\qquad = \dfrac{(g/GMW)_{NaCl}}{L_{solution}}$

PART 13—SOLUTIONS

$$\text{OEIOU:} \quad g_{NaCl} = M_{NaCl} \times L_{solution} \times GMW_{NaCl}$$
$$= (0.125 \text{ mole/L})(1.000 \text{ L})(58.44 \text{ g/mole})$$
$$= 7.31 \text{ g}$$

Question 16

What volume of 0.200 M $CaCl_2$ solution can be prepared from 1.11 g of solid $CaCl_2$?

Answer 16

0.0500 L (or 50.0 ml)

$$\text{OEIOU:} \quad L_{solution} = \frac{(g/GMW)_{CaCl_2}}{M_{CaCl_2}}$$

Question 17

A solution is formed by mixing 10.0 ml of ethyl alcohol (D = 0.789 g/ml) with 25.0 ml water. What is the molarity of the solution with respect to ethyl alcohol, C_2H_5OH?

Answer 17

4.89 M

$$M_{et} = \frac{n_{et}}{L_{solution}}$$

$$= \frac{(g/GMW)_{et}}{L_{et} + L_{H_2O}}$$

$$\text{OEIOU:} \quad M_{et} = \frac{(D \times ml/GMW)_{et}}{L_{et} + L_{H_2O}}$$

Question 18

How many grams of KOH must be dissolved to prepare 355 ml of 0.425 M potassium hydroxide solution?

Answer 18

8.46 g

PART 13–SOLUTIONS

Defining equation: $M = \dfrac{n_{KOH}}{L_{solution}}$

$= \dfrac{(g/GMW)_{KOH}}{L_{solution}}$

OEIOU: $g_{KOH} = M \times L_{solution} \times GMW_{KOH}$

Statement 6

The **molality** (m) of a solution, unlike its molarity, does not vary with temperature. Because molarity is defined in terms of the volume of solution, and volume changes somewhat as temperature changes, molarity will change to some extent if the temperature changes. Molality, however, is defined in terms of weights—the moles of solute per kilogram of solvent:

$$m_A = \dfrac{n_A}{kg_{solvent}}$$

Question 19

What is the molality of a solution formed by dissolving 0.300 mole of HNO_3 in 4.00 kg of water?

Answer 19

0.0750 m

OEIOU: $m_{HNO_3} = \dfrac{n_{HNO_3}}{kg_{H_2O}}$

Question 20

What is the molality of a solution formed by dissolving 11.9 g of potassium bromide in 500.0 g of water?

Answer 20

0.200 m

OEIOU: $m_{KBr} = \dfrac{(g/GMW)_{KBr}}{kg_{H_2O}}$

$= \dfrac{(11.9 \text{ g}/119 \text{ g/mole})}{0.5000 \text{ kg } H_2O}$

$= 0.200 \text{ mole } KBr/kg\ H_2O$

$= 0.200 \text{ m}$

Question 21

How many grams of naphthalene ($C_{10}H_8$) are required to prepare a 0.448 m solution of naphthalene in 100.0 ml of benzene (D = 0.894 g/ml)?

Answer 21

5.13 g

$$m_{nap} = \frac{(g/GMW)_{nap}}{ml_b \times D_b / \#g \text{ per kg}}$$

OEIOU: $$g_{nap} = \frac{m_{nap} \times ml_b \times D_b \times GMW_{nap}}{\#g \text{ per kg}}$$

Question 22

The density of 0.826 M NaBr is 1.063 g/ml at 25°C. What is the molality of the solution?

Answer 22

0.845 m

Defining equation: $m_{NaBr} = n_{NaBr}/kg_{H_2O}$

Substituting: $m_{NaBr} = n_{NaBr}/(kg_{total} - kg_{NaBr})$

For *each liter* of solution: $n_{NaBr} = M_{NaBr}$

Thus: $$m_{NaBr} = \frac{M_{NaBr}}{(kg_{tot} - kg_{NaBr})}$$

$$= \frac{M_{NaBr}}{(g_{tot} - g_{NaBr}) \times \#kg \text{ per g}}$$

$$= \frac{M_{NaBr}}{(ml \times D)_{solution} - (M \times GMW)_{NaBr} \times \#kg \text{ per g}}$$

OEIOU: $$m_{NaBr} = \frac{M_{NaBr} \times \#g \text{ per kg}}{(ml_{solution} \times D) - (M \times GMW)_{NaBr}}$$

Note: For problems of this type, when no volume is specified, one may assume a volume of one liter for convenience.

Statement 7

The **normality** of a solution is the number of gram equivalent weights per liter of solution:

$$N = \frac{\#GEW}{liters_{solution}}$$

To make the concept of GEW broad enough to encompass solutions of all types of solutes, it is defined as the gram molecular weight divided by the "**combining capacity**," n_c, of the solute:

$$GEW = \frac{GMW}{n_c}$$

Thus:

$$N = n_c M$$

The combining capacity of a substance depends on the reaction for which it is to be used. In general, we may make the following assumptions:

- **acids** n_c = number of reactive hydrogens per formula (usually x for acid of type $H_x Y$)
- **bases** n_c = number of reactive OH⁻ ions per formula (x for base of type $M(OH)_x$)
- **salts** n_c = total number of positive or negative charges per formula
- **oxidizing or reducing agents** n_c = number of electrons gained or lost per formula (Oxidizing and reducing agents will be considered in detail in Part 18.)

Strictly speaking, n_c should always be based on how the solute is to be used in a specific reaction, because the normality is an expression of concentration in terms of the units of "chemical reactivity" (GEWs) present per liter of solution. For example, an acid having more than one reactive hydrogen, $H_x Y$, may be used to supply only one hydrogen in one specific reaction, while in another it might supply more than one; thus its value of n_c may vary from $n_c = 1$ up to $n_c = x$, depending on the reaction. The preceding guidelines assume that the solute is being used to its maximum chemical capacity.

Question 23

What is the combining capacity of:

a. H_2SO_4?

b. $Ca(OH)_2$?

c. NaCl ?

Answer 23

a. $n_c = 2$

b. $n_c = 2$

c. $n_c = 1$

Question 24

When $KMnO_4$ acts as an oxidizing agent in acid solutions, we may show what happens to the MnO_4^- ion by the equation:

$$5e^- + MnO_4^- + 8H^+ \rightarrow Mn^{2+} + 4H_2O$$

where e^- represents an electron. What is the combining capacity of $KMnO_4$?

Answer 24

$$n_c = 5$$

For each MnO_4^- ion (thus for each $KMnO_4$), five electrons are gained.

Question 25

Calculate the GEW of:

a. H_3PO_4

b. NaOH

Answer 25

a. 32.7 g/GEW

$$GEW = GMW/n_c = \frac{98.0 \text{ g/mole}}{3 \text{ GEW/mole}}$$

$$= 32.7 \text{ g/GEW}$$

We are assuming that the H_3PO_4 is to be used in a reaction requiring all three hydrogens.

b. 40.0 g/GEW

$$\text{OEIOU: } GEW = GMW/n_c = \frac{40.0 \text{ g/mole}}{1 \text{ GEW/mole}}$$

$$= 40.0 \text{ g/GEW}$$

Question 26

What is the GEW of:

a. $CaCl_2$

b. $KMnO_4$ reacting as an oxidizing agent in acid solution. (See Question 24.)

Answer 26

a. 55.5 g/GEW

OEIOU: GEW = GMW/n_c = $\dfrac{111 \text{ g/mole}}{2 \text{ GEW/mole}}$

= 55.5 g/GEW

b. 31.6 g/GEW

(n_c = 5 GEW/mole)

Question 27

What is the normality of a solution formed by dissolving 55.0 g of $CaCl_2$ in 500 ml of water?

Answer 27

1.98 N

Defining equation: N = $\dfrac{\#GEW}{\#L}$

Substituting: N = $\dfrac{\#g/GEW}{\#L}$

OEIOU: N = $\dfrac{\#g/(GMW/n_o)}{\#L}$

= $\dfrac{55.0 \text{g}/(111 \text{ g/mole}/2\text{GEW/mole})}{0.500 \text{ L}}$

= 1.98 GEW/L

= 1.98 N

Question 28

What is the normality of a liter of solution containing:

a. 32.7 g H_2SO_4

b. 15.4 g $Ba(OH)_2$

Answer 28

a. 0.667 N

b. 0.180 N

Question 29

What is the normality of:

a. a 0.100 M H_2SO_4 solution?

b. a 0.100 M NaOH solution?

Answer 29

a. 0.200 N

$$N = n_c M$$

$$= (2 \text{GEW/mole}) \times (0.100 \text{ mole/L})$$

$$= 0.200 \text{ GEW/L}$$

b. 0.100 N

Statement 8

It is often convenient to prepare a solution by **diluting** a more concentrated solution containing the desired solute, rather than dissolving a weighed amount of solute. In deciding how much of the more concentrated solution is needed, we may represent the dilution process by the equation:

$$\text{solute (solution, } M_1) \rightarrow \text{solute (solution, } M_2)$$

We may then write a **stoichiometric statement** for the process:

$$\# \text{moles, } M_1 = \# \text{moles, } M_2$$

From the equation $M = \# \text{moles}/L_{solution}$, we know that $\# \text{moles} = M \times L_{solution}$. Therefore, rearranging, we get:

$$M_1 \times L_1 = M_2 \times L_2$$

This is the fundamental equation used in dilution problems. Any type of volume unit may be used, provided that the **same units** are used on both sides of the equation. Similar reasoning when N rather than M is used leads to a similar expression:

$$N_1 \times L_1 = N_2 \times L_2$$

Question 30

A 12 M HCl solution is to be used to prepare 0.100 L of 8.3 M HCl. Show a stoichiometric statement for the dilution process.

Answer 30

$$\# \text{moles, 8.3 M} = \# \text{moles, 12 M}$$

Question 31

How much 12 M HCl would be needed to prepare 0.100 L of 8.3 M HCl?

Answer 31

0.069 L (or 69 ml)

SS: # moles, 8.3 M = # moles, 12 M

Substituting: $M_1 \times L_1 = M_2 \times L_2$

OEIOU: $L_2 = \dfrac{M_1 \times L_1}{M_2}$

$= \dfrac{8.3 \text{ M} \times 0.100 \text{ L}}{12 \text{M}}$

$= 0.069$ L

Question 32

How much water must be added to the 12 M HCl in Question 31 to give 0.100 liter of 8.3 M HCl?

Answer 32

0.031 L (or 31 ml)

The solution was formed by mixing some 12 M solution with some water; thus:

Defining equation: $L_{8.3 \text{ M}} = L_{12 \text{ M}} + L_{H_2O}$

OEIOU: $L_{H_2O} = L_{8.3 \text{ M}} - L_{12 \text{ M}}$

$= 0.100 \text{ L} - 0.069 \text{ L}$

$= 0.031$ L

Question 33

How many ml of 12 M sulfuric acid and how many ml of water would be needed to prepare 250 ml of 2.5 M H_2SO_4 solution?

Answer 33

52 ml H_2SO_4 and 198 ml water

OEIOU: $ml_{12\,M} = \dfrac{M_{2.5\,M} \times ml_{2.5\,M}}{M_{12\,M}}$

OEIOU: $ml_{H_2O} = ml_{2.5\,M} - ml_{12\,M}$

Question 34

How many ml of 0.680 M KBr could be made from 50.0 ml of 1.00 M KBr?

Answer 34

73.5 ml

OEIOU: $ml_{0.68\,M} = M_{1\,M} \times L_{1\,M}/M_{0.68\,M}$

Question 35

What is the molarity of the resulting solution when 30.0 ml of 2.0 M ammonia solution is diluted to 50.0 ml?

Answer 35

1.2 M

OEIOU: $M_{final} = M_{2.0\,M} \times ml_{2.0\,M}/ml_{final}$

Question 36

What is the molarity of the resulting solution when 2.00 L of 6.00 M HNO_3 is diluted with 1.00 L of water?

Answer 36

4.0 M

Defining equation: $M_f \times L_f = M_{6.0} \times L_{6.0\,M}$

Substituting: $M_f (L_{6.0} + L_{H_2O}) = M_{6.0} \times L_{6.0}$

OEIOU: $M_f = \dfrac{M_{6.0\,M} \times L_{6.0\,M}}{L_{6.0\,M} + L_{H_2O}}$

Question 37

What is the normality of a solution formed by mixing 50.0 ml of 0.100 N KOH with 75.0 ml of water?

Answer 37

0.0400 N

Defining equation: $N_1 \times ml_1 = N_2 \times ml_2$

$$N_2 = \frac{N_1 \times ml_1}{ml_2}$$

OEIOU: $N_2 = \dfrac{N_1 \times ml_1}{ml_1 + ml_{H_2O}}$

$= \dfrac{0.100 \text{ N} \times 50.0 \text{ ml}}{50.0 \text{ ml} + 75.0 \text{ ml}}$

$= 0.0400 \text{ N}$

Question 38

How much 0.100 M H_2SO_4 is needed to form 250 ml of 0.0400 N H_2SO_4?

Answer 38

50.0 ml

Defining equation: $N_1 \times ml_1 = N_2 \times ml_2$

Substituting: $n_c M_1 \times ml_1 = N_2 \times ml_2$

OEIOU: $ml_1 = \dfrac{N_2 \times ml_2}{n_c M_1}$

Statement 9

Henry's Law. Though pressure changes have a negligible effect on the solubility of solid and liquid solutes, the solubility of a **gas** in a liquid solvent at moderate pressure is proportional to its partial pressure above the solution:

$$C_A = kP_A$$

where C_A is the concentration of the gaseous solute in solution, k is a proportionality constant, and P_A is the partial pressure of the solute.

Question 39

The Henry's Law constant for O_2 in water at $20°C$ is 1.38×10^{-3} mole/L-atm. What is the solubility of oxygen in a solution when its partial pressure is 0.30 atm?

Answer 39

4.1×10^{-4} mole/L

$$C_{O_2} = k \times P_{O_2}$$

$$= (1.38 \times 10^{-3} \text{ mole/L-atm}) \times (0.30 \text{ atm})$$

$$= 4.1 \times 10^{-4} \text{ mole/L}$$

Question 40

The solubility of oxygen gas in water at 1.00 atm and $20°C$ is 4.42×10^{-2} gm/L. What is its solubility at 3.50 atm?

Answer 40

0.155 g/L

Since k has the same value for both pressures:

$$\frac{C_{3.5 \text{ atm}}}{P_{3.5 \text{ atm}}} = \frac{C_{1 \text{ atm}}}{P_{1 \text{ atm}}}$$

OEIOU: $C_{3.5} = C_1 \times \dfrac{P_{3.5}}{P_1}$

Statement 10

For **ideal solutions** (those for which interactions between solute and solvent are negligible), **Raoult's Law** states that the partial pressure, P_1, of a component of a solution equals the mole fraction of the component, X_1, times its normal vapor pressure when pure, $P_1^°$:

$$P_1 = X_1 P_1^°$$

It follows, then, since X_1 is always less than one, that P_1 is always less than $P_1^°$. When the component is the solvent, we may say that the **vapor pressure** of the solvent has been **lowered** by the solute. Mathematically:

$$P_1 = (1 - X_2) P_1^°$$

or:

$$P_1^\circ - P_1 = \Delta P_1 = X_2 P_1^\circ$$

where X_2 is the mole fraction of the solute.

Question 41

What is the partial pressure of water over a solution formed by dissolving 2.00 moles of methyl alcohol (CH_3OH) in 5.00 moles of water at 25°C? (The vapor pressure of water at 25°C is 23.8 mm Hg.)

Answer 41

17.0 mm Hg

Defining equation: $P_{H_2O} = X_{H_2O} P^\circ_{H_2O}$

OEIOU: $P_{H_2O} = \dfrac{n_{H_2O}}{n_{H_2O} + n_{Me}} P^\circ_{H_2O}$

Question 42

What is the decrease in vapor pressure produced by dissolving 18.0 g of glucose, $C_6H_{12}O_6$, in 90.0 g of water at 25°C? (The vapor pressure of water at 25°C is 23.8 torr.)

Answer 42

0.466 torr

OEIOU: $\Delta P_{H_2O} = \dfrac{(g/GMW)_{Gl}}{(g/GMW)_{Gl} + (g/GMW)_{H_2O}} \times P^\circ_{H_2O}$

Statement 11

Lowering the vapor pressure of a solvent also lowers its **freezing point**, T_f, and raises its **boiling point**, T_b. Mathematically:

$$\Delta T_f = K_f m \quad \text{and} \quad \Delta T_b = K_b m$$

where K_f and K_b are proportionality constants called the "freezing point constant" and the "boiling point constant," respectively, of the solvent, and m is the molality of the solute.

Question 43

To what extent is the freezing point of water lowered when 18.0 g of glucose is added to 90.0 g of water? (K_f for water is 1.86°C kg/mole.)

Answer 43

2.07°C

Defining equation: $\Delta T_f = K_f m$

Substituting: $\Delta T_f = K_f \times \dfrac{(\text{g/GMW})_{Gl}}{\text{kg}_{H_2O}}$

$= (1.86°\text{C kg/mole}) \dfrac{(18.0 \text{ g}/180 \text{ g/mole})}{(0.0900 \text{ kg})}$

$= 2.07°\text{C}$

Question 44

How much is the boiling point of water raised by dissolving 18.0 g of glucose in 90.0 g of water? (K_b for water is 0.52°C kg/mole.)

Answer 44

0.58°C

OEIOU: $\Delta T_b = K_b \times \dfrac{(\text{g/GMW})_{Gl}}{\text{kg}_{H_2O}}$

Question 45

1.30 g of a certain organic compound lowered the freezing point of 50.0 g of water by 0.81°C. What is the molecular weight of the compound? (K_f for water is 1.86°C kg/mole.)

Answer 45

60 g/mole

Defining equation: $\Delta T_f = K_f m$

$= K_f \times \dfrac{(\text{g/GMW})_{cmpd}}{\text{kg}_{H_2O}}$

OEIOU: $\text{GMW}_{cmpd} = \dfrac{K_f \times g_{cmpd}}{kg_{H_2O} \times T_f}$

Question 46

What is the freezing point of a solution formed by adding 4.50 g of glucose to 90.0 g of water at 1 atm? (K_f for water is 1.86°C kg/mole.)

Answer 46

−0.52°C

Defining equation: $T_{f_2} = T_{f_1} - \Delta T_f$

OEIOU: $T_{f_2} = T_{f_1} - K_f \dfrac{(\#g/GMW)_{Gl}}{kg_{H_2O}}$

$= 0.00°C - (1.86°C\ kg/mole)\dfrac{(4.50\ g/180\ g/mole)}{0.090\ kg}$

$= 0.00°C - 0.52°C$

$= -0.52°C$

Question 47

1.30 g of urea, CH_4N_2O, is dissolved in 50.0 g of camphor. What is ΔT_f for camphor if its K_f is 40.0°C kg/mole?

Answer 47

17.3°C

OEIOU: $\Delta T_f = K_f \times \dfrac{(g/GMW)_{Ur}}{kg_{cam}}$

$= 40.0°C\ kg/mole\ \dfrac{(1.30\ g/60.0\ g/mole)}{0.050\ kg}$

$= 17.3°C$

(Compare with the small change obtained with water in Question 41.)

Question 48

"Permanent" antifreeze is ethylene glycol, $C_2H_6O_2$. How many grams of antifreeze would be needed per liter of water to protect an automobile radiator against temperatures as low as $-10.0°C$? ($D_{H_2O} = 1.00$ gm/ml; K_f for water is $1.86°C$ kg/mole.)

Answer 48

333 g

Defining equation: $\Delta T_f = K_f m$

Substituting: $T_{f_1} - T_{f_2} = K_f \times \dfrac{(g/GMW)_{Ant}}{kg_{H_2O}}$

or

$$g_{Ant} = (T_{f_1} - T_{f_2}) \times \dfrac{kg_{H_2O} \times GMW_{Ant}}{K_f}$$

OEIOU: $g_{Ant} = (T_{f_1} - T_{f_2}) \times \dfrac{(ml \times D)_{H_2O} \,(\# kg \text{ per g}) \times GMW_{Ant}}{K_f}$

Statement 12

The changes in P_{vap}, T_f, and T_b produced by dissolving a solute in a solvent are dependent on the **number of moles of particles** of solute present, whether atoms, ions, or molecules. Solutes that **ionize** in solution produce changes proportionate to the number of ions furnished by them. We would expect a mole of NaCl to produce two moles of ions, for example. To take into account the ionization effect of solutes that are electrolytes, we may alter the equations given in Statement 11:

$$\Delta T_f = n_i K_f m \quad \text{and} \quad \Delta T_b = n_i K_b m$$

where n_i (sometimes called the **van't Hoff factor**) is the number of moles of ions produced per mole of solute. The number predicted from formulas is actually the limiting value of n_i; at ordinary concentrations interactions among ions give values smaller than this.

Question 49

What is the boiling point elevation produced by dissolving 29.2 g of NaCl in a liter of water? (K_f for water is $0.52°C$ kg/mole.)

Answer 49

0.52°C

OEIOU: $\Delta T_f = n_i K_f \times \dfrac{(g/GMW)_{NaCl}}{(ml \times D \times kg/g)_{H_2O}}$

Question 50

The freezing point of a 0.10 m $NaHCO_3$ solution is $-0.38°C$. How many ions are produced per formula of sodium bicarbonate when it is dissolved in water? (K_f for water is $1.86°C$ kg/mole.)

Answer 50

2.0

OEIOU: $n_i = \dfrac{T_{f_1} - T_{f_2}}{K_f \times m_{NaCl}}$

$= \dfrac{0.00°C - (-0.38°C)}{1.86°C \text{ kg/mole} \times 0.10 \text{ m}}$

$= 2.0$

Statement 13

Osmosis is the passing of solvent molecules through a semipermeable membrane separating two solutions of different concentration. The **osmotic pressure** Π, is the pressure required to prevent any net transfer of pure solvent into a solution across a semipermeable membrane.

Π is proportional to the molarity of the solute. Mathematically:

$$\Pi V = nRT \quad \text{or} \quad \Pi = MRT$$

where V is the volume of solution, R is the gas constant (0.0821 L-atm/mole°K), T is the absolute temperature, and M is the molarity of the solute.

Question 51

What is the osmotic pressure of a solution that is 0.10 M in glucose, $C_6H_{12}O_6$, at 298°K?

Answer 51

2.45 atm

OEIOU: $\Pi = MRT$

$= (0.10\ M)(0.0821\ \text{L-atm/mole}°K)(298°K)$

$= 2.45\ \text{atm}$

Question 52

Blood has an osmotic pressure of about 7.7 atm. What is the total number of moles of dissolved constituents per liter of blood? (Body temperature = 98.6°F.)

Answer 52

0.30 M

OEIOU: $M = \Pi/RT$

$= \dfrac{7.7\ \text{atm}}{(0.0821\ \text{L-atm/mole}°K)(310°K)}$

$= 0.30\ M$

Note that this value actually represents the total moles of *particles* per liter, both ions and molecules, because blood contains both electrolytes and nonelectrolytes.

TESTING YOUR MASTERY—Part 13 Solutions

Question 53

Which component in each of these mixtures is considered to be the solvent?

a. 100 ml H_2O and 800 ml ethyl alcohol

b. 100 g H_2O and 10 g KCl

c. 100 g H_2O and 10 g Na_2CO_3 and 30 g methyl alcohol

d. 100 g H_2O and 10 g KOH and 300 ml ethylene glycol

e. 1.6 g CO and 100 ml ether

Answer 53

a. ethyl alcohol

b. water

c. water

d. ethylene glycol

e. ether

Question 54

If 3.20 g of methyl alcohol, CH_4O (density - 0.793 g/ml) is mixed with 180.0 g of water (density - 1.00 g/ml), what is:

a. the % by weight of methyl alcohol?

b. the % by volume of methyl alcohol?

c. the mole fraction of methyl alcohol?

Answer 54

a. 1.75% by weight

$$\text{OEIOU: } \%_{Me} = \frac{g_{Me}}{g_{Me} + g_{H_2O}} \times 100$$

b. 2.19% by volume

$$\text{OEIOU: vol. } \%_{Me} = \frac{(g/D)_{Me}}{(g/D)_{Me} + (g/D)_{H_2O}} \times 100$$

c. 0.00990 mole fraction

$$\text{OEIOU: } X_{Me} = \frac{(g/GMW)_{Me}}{(g/GMW)_{Me} + (g/GMW)_{H_2O}}$$

Question 55

A solution is prepared by dissolving 10.6 g of Na_2CO_3 in 90.0 ml of water (density - 1.00 g/ml). Calculate the following:

a. the mole fraction of Na_2CO_3.

b. the molarity of Na_2CO_3.

c. the molality of Na_2CO_3.

Answer 55

a. 0.0196 mole fraction Na_2CO_3

OEIOU: $X_{Na_2CO_3} = \dfrac{(g/GMW)_{Na_2CO_3}}{(g/GMW)_{Na_2CO_3} + \left(\dfrac{D \times ml}{GMW}\right)_{H_2O}}$

b. 1.11 M

OEIOU: $M_{Na_2CO_3} \approx \dfrac{(g/GMW)_{Na_2CO_3}}{L_{H_2O}}$

c. 1.11 M

OEIOU: $M_{Na_2CO_3} = \dfrac{(g/GMW)_{Na_2CO_3}}{L_{H_2O} \times \#kg\, H_2O\text{ per L}}$

Question 56

For the following solutions, find both the molarity and normality of the solute.

a. 10.0 g KOH in 100.0 ml H_2O

b. 10.0 g $Ba(OH)_2$ in 100.0 ml H_2O

Answer 56

a. 1.78 M and 1.78 N

OEIOU: $M_{base} \approx \dfrac{(g/GMW)_{base}}{L_{H_2O}}$

OEIOU: $N_{base} \approx n_{c,base} \times \dfrac{(g/GMW_{base})}{L_{H_2O}}$

b. 0.583 and 1.17 N

Same OEIOU in each case as for KOH, but $n_c = 1$ for KOH, whereas $n_c = 2$ for $Ba(OH)_2$.

Question 57

How many moles of H_2SO_4 are needed to prepare 100.0 ml of the following solutions:

a. 0.400 M H_2SO_4

b. 0.400 N H_2SO_4

Answer 57

a. 0.0400 mole

OEIOU: # moles acid = $M_{acid} \times L_{solution}$

b. 0.0200 mole

OEIOU: # moles acid = $\dfrac{N_{acid} \times L_{solution}}{n_{c,\,acid}}$

Question 58

What is the normality of the following solutions:

a. 0.100 M KCl

b. 0.100 M $KMnO_4$ used as an oxidizing agent in a basic solution, in which the following equation shows what happens to the MnO_4 ion:

$$3e^- + MnO_4^- + 2H_2O \rightarrow MnO_2 + 4OH^-$$

Answer 58

a. 0.100 N

b. 0.300 N

OEIOU: $N = n_c \times M$

Same OEIOU for KCl and $KMnO_4$, but $n_c = 1$ for KCl, whereas $n_c = 3$ for $KMnO_4$.

Question 59

How many grams of solute are needed to prepare 200.0 ml of the following solutions in water?

a. 0.100 M NaBr

b. 0.100 mole fraction NaBr

Answer 59

a. 2.06 g NaBr

OEIOU: $\#g_{NaBr} = GMW_{NaBr} \times L_{solution} \times M$

b. 127 g NaBr

OEIOU: $g_{NaBr} = \dfrac{X_{NaBr} \times (D \times ml)_{H_2O} \times GMW_{NaBr}}{(1 - X_{NaBr}) \times GMW_{H_2O}}$

Question 60

What volumes of the following solutions could be prepared from 12.1 g of $MgSO_4$?

a. 0.500 M $MgSO_4$

b. 0.500 N $MgSO_4$

Answer 60

a. 0.201 L

OEIOU: $\#L_{solution} = \dfrac{(g/GMW)_{MgSO_4}}{M}$

b. 0.402 L

OEIOU: $\#L_{solution} = n_{c,MgSO_4} \times \dfrac{(g/GMW)_{MgSO_4}}{N}$

Question 61

The density of a 2.32 M solution of H_2SO_4 is 1.139 g/ml. What is the molality of the solution?

Answer 61

2.54 m

OEIOU: $m_{H_2SO_4} = \dfrac{M_{H_2SO_4} \times \#g \text{ per kg}}{(ml_{solution} \times D) - (M \times GMW)_{H_2SO_4}}$

Question 62

How many ml of 11.6 M HCl solution are needed to prepare 750 ml of 6.41 M HCl?

Answer 62

414 ml

OEIOU: $ml_1 = \dfrac{ml_2 \times M_2}{M_1}$

Question 63

How much 5.00 M NaOH could be prepared from 175 ml of 7.50 M NaOH solution?

Answer 63

262 ml

OEIOU: $ml_2 = \dfrac{M_1 \times ml_1}{M_2}$

Question 64

a. How much 5.00 M H_2SO_4 solution could be prepared from 100.0 ml of 12.0 M H_2SO_4?

b. How much 5.00 N H_2SO_4 solution could be prepared from the same 100.0 ml of 12.0 M H_2SO_4?

Answer 64

a. 240 ml

OEIOU: $ml_2 = \dfrac{M_1 \times ml_1}{M_2}$

b. 480 ml

OEIOU: $ml_2 = \dfrac{M_1 \times ml_1}{N_2} \times n_{c, H_2SO_4}$

Question 65

What is the normality of a solution formed by mixing:

a. 50.5 ml of 3.00 M $CaCl_2$ and 505 ml of H_2O?

b. 50.5 ml of 3.00 N $CaCl_2$ and 505 ml of H_2O?

Answer 65

a. 0.273 N

OEIOU: $N_2 = \dfrac{M_1 \times ml_1}{ml_{H_2O} + ml_1} \times n_{c, CaCl_2}$

b. 0.273 N

OEIOU: $N_2 = \dfrac{N_1 \times ml_1}{ml_{H_2O} + ml_1}$

Question 66

How many ml of 6.00 M HNO_3 and how many ml of water are needed to prepare 125 ml of 2.50 M HNO_3?

Answer 66

52.1 ml HNO_3 and 72.9 ml H_2O

OEIOU: $ml_1 = \dfrac{M_2 \times ml_2}{M_1}$

OEIOU: $ml_{H_2O} = ml_2 - ml_1$

Question 67

A 56.0 per cent solution of HNO_3 has a density of 1.35 g/ml. How much of this solution is required to prepare 125 ml of 2.50 M HNO_3?

Answer 67

14.6 ml

OEIOU: $ml_1 = \dfrac{M_2 \times L_2 \times GMW_{HNO_3}}{D_1 \times (\% HNO_3/100)}$

Question 68

What is the molarity of the resulting $Hg(NO_3)_2$ solution when 20.0 ml of 0.800 M $Hg(NO_3)_2$ is mixed with the following:

a. 180.0 ml of water

b. 180.0 ml of 6.0 M HNO_3

Answer 68

a. 0.0800 M $Hg(NO_3)_2$

OEIOU: $M_2 = M_1 = \dfrac{M_1 \times ml_1}{ml_1 + ml_{H_2O}}$

b. 0.0800 M $Hg(NO_3)_2$

$$\text{OEIOU: } M_2 = \frac{M_1 \times ml_1}{ml_1 + ml_{HNO_3}}$$

Notice that the M of $Hg(NO_3)_2$ is the same whether water or HNO_3 solution is used for dilution.

Question 69

The solubility of O_2 gas in water is 2.76×10^{-3} mole/L at 20°C and 2.00 atm. What should be its solubility at 4.68 atm?

Answer 69

6.45×10^{-3} mole/L

$$\text{OEIOU: } C_{P_2} = C_{P_1} \times \frac{P_2}{P_1}$$

Question 70

The solubility of O_2 gas in water is 4.42×10^{-2} g/L at 1.00 atm and 20°C. At what pressure would 0.100 g of O_2 dissolve in a liter of water?

Answer 70

2.26 atm

$$\text{OEIOU: } P_2 = \frac{C_{P_2}}{C_{P_1}} \times P_1$$

Question 71

The vapor pressure of water at 20°C is 17.5 torr. What is the partial pressure of water over a solution formed by dissolving 0.466 mole of ethyl alcohol in 3.00 moles of water at 20°C?

Answer 71

15.1 torr

$$\text{OEIOU: } P_{H_2O} = \frac{n_{H_2O}}{n_{H_2O} + n_{Et}} \times P^o_{H_2O}$$

Question 72

K_f for water is 1.86°C kg/mole, and K_b is 0.52°C kg/mole. When 17.1 g of sucrose, $C_{12}H_{22}O_{11}$, is dissolved in 500.0 g of water, what is:

a. the change in the freezing point?

b. the change in the boiling point?

Answer 72

a. 0.186°C

$$\text{OEIOU:} \quad \Delta T_f = K_f \times \frac{(g/GMW)_{suc}}{kg_{H_2O}}$$

b. 0.052°C

$$\text{OEIOU:} \quad \Delta T_b = K_b \times \frac{(g/GMW)_{suc}}{kg_{H_2O}}$$

Question 73

The vapor pressure of pure water at 22°C is 19.8 torr. When 30.0 g of fructose, $C_6H_{12}O_6$, is dissolved in 100.0 g of water, what is:

a. the change in the vapor pressure of the water?

b. the osmotic pressure of the solution?

Answer 73

a. 0.574 torr

$$\text{OEIOU:} \quad \Delta P_{H_2O} = (g/GMW)_{fr} + \frac{(g/GMW)_{fr}}{(g/GMW)_{H_2O}} \times P^\circ_{H_2O}$$

b. 40.2 atm (or 3.05 × 10⁴ torr)

$$\text{OEIOU:} \quad \Pi = \frac{(g/GMW)_{fr} \, RT}{\#kg_{H_2O} \times (L/kg)}$$

Question 74

If 50.0 g of ethylene glycol, $C_2H_6O_2$, is dissolved in 245 g of water, what is:

a. the freezing point of the resulting solution?
b. the boiling point of the resulting solution?

$(K_{f,H_2O} = 1.86°C \text{ kg/mole})$

$(K_{b,H_2O} = 0.520°C \text{ kg/mole})$

Answer 74

a. $-6.12°C$

OEIOU: $T_{f_2} = T_{f_1} - K_f \dfrac{(g/GMW)_{Et}}{kg_{H_2O}}$

b. $+101.71°C$

OEIOU: $T_{b_2} = T_{b_1} + K_b \dfrac{(g/GMW)_{Et}}{kg_{H_2O}}$

Question 75

A 2.00-g sample of an unknown organic compound produced a 2.50°C lowering of the freezing point of 100.0 g of camphor. What is the molecular weight of the compound? (K_f for camphor is 40.0°C kg/mole)

Answer 75

320 g/mole

OEIOU: $GMW_{cmpd} = \dfrac{K_f \times g_{cmpd}}{Kg_{cam} \times \Delta T_f}$

Question 76

The boiling point of a solution formed by dissolving 11.9 g of KBr in 100.0 g of water is 101.0°C. Approximately how many ions are formed per formula of KBr when it is dissolved in water?

Answer 76

$n_i \approx 2$

OEIOU: $n_i = \dfrac{(T_2 - T_1) \times kg_{H_2O} \times GMW_{KBr}}{K_b \times g_{KBr}}$

Question 77

The freezing point of a 0.0100 m solution of NaCl is −0.036°C, while a 0.0100 m $MgSO_4$ solution freezes at −0.028°C. In which solution is there a stronger interaction among ions in solution?

Answer 77

In the $MgSO_4$ solution.

From these data, for NaCl $n_i = 1.94$, and for $MgSO_4$ $n_i = 1.51$. Thus the Mg^{2+} and SO_4^{2-} ions are not acting independently, and therefore must be interacting more strongly.

Question 78

A certain aqueous solution of urea has an osmotic pressure of 1.76 atm at 25°C. What is the molarity of the solution?

Answer 78

0.0719 M

OEIOU: $M = \Pi/RT$

PART 14
KINETICS

Statement 1

Chemical kinetics is the area of chemistry that deals with **rates** of chemical reactions. By rate of reaction we mean the change in concentration of reactants or products per unit time. For a generalized one-step reaction (that is, a reaction resulting from a single collision between reactant particles),

$$aA + bB \rightarrow \text{products}$$

we find that the **rate of reaction** is given by the so-called **Law of Mass Action:***

$$\text{rate} = k[A]^a[B]^b$$

where k is a proportionality constant called the **rate constant**, a and b are the coefficients in the balanced equation, and $[A]$ and $[B]$ are the concentrations (most commonly molarities) of the reactants. This type of equation is usually called a **rate equation**.

---

*Most texts reserve this title for what we will call the "Law of Chemical Equilibrium" in Part 15. Historically, the rate expression rather than the equilibrium expression was called the Law of Mass Action, and the Law of Chemical Equilibrium was derived from it.

Question 1

Write the rate equation for this reaction:

$$H_2 + I_2 \rightarrow 2HI$$

Answer 1

rate = $k[H_2][I_2]$

Question 2

Write the rate equation for this reaction:

$$2NO_2 \rightarrow 2NO + O_2$$

Answer 2

rate = $k[NO_2]^2$

Statement 2

Many reactions occur in several steps rather than one step, often leading to more complex rate equations. Whatever the case, for the rate equation

$$\text{rate} = k[A]^a[B]^b$$

the sum of the superscripts a and b is called the **order of the reaction** (first order, second order, and so forth):

$$\text{order} = a + b$$

In addition, the reaction is said to be "ath order in A" and "bth order in B."

Question 3

For $N_2O_4 \rightarrow 2NO_2$ the rate equation is: rate = $k[N_2O_4]$. What is the order of the reaction?

Answer 3

first order

The only superscript in the expression is the implied "1" on the $[N_2O_4]$ term. Thus the reaction must be *first order*.

Question 4

$H_2 + I_2 \rightarrow 2HI$

rate = $k[H_2][I_2]$

What is:

a. the overall order of reaction?
b. the order with respect to H_2?
c. the order with respect to I_2?

Answer 4

a. second order

$$\text{order} = a + b = 1 + 1 = 2$$

b. first order with respect to H_2

c. first order with respect to I_2

Question 5

The reaction $2NO + O_2 \rightarrow 2NO_2$ has the rate equation:

$$\text{rate} = k[NO]^2[O_2]$$

What is:

a. the overall order?

b. the order with respect to NO?

c. the order with respect to O_2?

Answer 5

a. third order overall

$$\text{order} = a + b = 2 + 1 = 3$$

b. second order in NO

c. first order in O_2

Question 6

For the reaction $2H_2 + O_2 \rightarrow 2H_2O$, the rate expression is:

$$\text{rate} = k[H_2][O_2]^{\frac{4}{3}}$$

a. What is the overall order?

b. Is this a one-step reaction?

Answer 6

a. $\frac{7}{3}$ or 2.33

b. No. If it were, the rate expression would be:

$$\text{rate} = k[H_2]^2[O_2]$$

Question 7

What are the units of a first-order rate constant if the time is measured in minutes?

Answer 7

min^{-1} (or 1/min)

$$rate = k[A]$$
$$k = rate/[A]$$
$$= \left(\frac{mole/L}{min}\right) / (mole/L)$$
$$= min^{-1}$$

Question 8

What are the units of a second-order rate constant if the time is measured in minutes?

Answer 8

$L\ mole^{-1}\ sec^{-1}$

$$rate = k[A][B]$$
$$k = rate/[A][B]$$
$$= \left(\frac{mole/L}{min}\right) / (mole/L)(mole/L)$$
$$= L\ mole^{-1}\ min^{-1}$$

Statement 3

The **half-life**, $t_{\frac{1}{2}}$, of a reaction is the time taken for half of a reactant to be consumed. It is a useful concept in that equal fractions of a reactant are consumed in equal periods of time.

Derivations of $t_{\frac{1}{2}}$ lead to different expressions for reactions of different orders.

For a first order reaction: $A \rightarrow$ products

$$t_{\frac{1}{2}} = 0.693/k$$

For a second order reaction: $2A \rightarrow$ products

$$t_{\frac{1}{2}} = 1/k[A]_o$$

where $[A]_o$ is the initial molarity of A.

The $t_{\frac{1}{2}}$ expression becomes more complex for other types of reactions.

Question 9

For the reaction $N_2O_5 \rightarrow 2NO_2 + \frac{1}{2}O_2$

$k = 4.0 \times 10^{-5}$ sec^{-1} at 27°C.

If the initial concentration of N_2O_5 is 4.0×10^{-2} mole/liter, what is the half-life of the reaction?

Answer 9

1.7×10^4 sec

Judging from units on k, this must be a first order reaction. Therefore:

$$t_{\frac{1}{2}} = 0.693/k$$

Question 10

The rate constant for the reaction $2NOCl \rightarrow 2NO + Cl_2$ is 2.8×10^{-5} L mole^{-1} sec^{-1} at 27°C. If the initial concentration of NOCl is 0.040 mole/L, after how many seconds would the concentration be 0.020 mole/L?

Answer 10

8.9×10^5 sec

From the units on k, this must be a second order reaction. Therefore:

$$t_{\frac{1}{2}} = 1/k[NOCl]_o$$

Statement 4

The rate of a reaction usually increases as temperature increases. Heating a reaction mixture supplies energy, increasing the average energy of reactant molecules so that more of them have the necessary **activation energy**, E_a, to undergo reaction.

The effect on k of changing T is given by the **Arrhenius equation**:

$$k = Ae^{-E_a/RT}$$

where A is a constant for the specific reaction, e is the natural log base, and R is the gas constant (1.99 cal/mole°K).

Question 11

For the reaction $2NOCl \rightarrow 2NO + Cl_2$, $E_a = 24$ kcal and A is 8.9×10^{12} L mole^{-1} sec^{-1} at 27°C. Calculate k for this reaction.

Answer 11

3.1×10^{-5} L mole^{-1} sec^{-1}

Defining equation: $k = Ae^{-E_a/RT}$

Thus: $\ln k = \ln A + (-E_a/RT)$

or $\quad 2.303 \log k = 2.303 \log A - E_a/RT$

$$\log k = \log A - \frac{E_a}{2.303\, RT}$$

OEIOU: $k = \text{antilog}\left(\log A - \dfrac{E_a}{2.303\, RT}\right)$

$\qquad = \text{antilog}\left(12.95 - \dfrac{2.4 \times 10^4 \text{ cal}}{(2.303)(1.99 \text{ cal/mole°K})(300°K)}\right)$

$\qquad = \text{antilog}\, (12.95 - 17.46) = \text{antilog}\,(-4.51)$

$\qquad = 3.1 \times 10^{-5}$ L mole^{-1} sec^{-1}

Question 12

The activation energy for the reaction $2N_2O_5 \rightarrow 4NO_2 + O_2$ is 25 kcal, and k is 4.0×10^{-5} sec^{-1} at 27°C. What will be the value of k if the temperature is increased to 37°C?

Answer 12

1.5×10^{-4} sec^{-1}

$$k_1 = Ae^{-E_a/RT_1} \quad \text{(at 27°C)}$$

$$k_2 = Ae^{-E_a/RT_2} \quad \text{(at 37°C)}$$

Thus: $\dfrac{k_2}{k_1} = \dfrac{Ae^{-E_a/RT_2}}{Ae^{-E_a/RT_1}}$

$\qquad\qquad = e^{\dfrac{E_a\,(T_2 - T_1)}{R\, T_1 T_2}}$

$$\log \frac{k_2}{k_1} = \frac{E_a(T_2 - T_1)}{2.303\,R.T_1T_2}$$

OEIOU: $k_2 = \text{antilog}\left[\dfrac{E_a(T_2 - T_1)}{2.303\,R\,T_1 T_2} + \log k_1\right]$

$\qquad\quad = \text{antilog}\,(0.587 - 4.398)$

$\qquad\quad = \text{antilog}\,(-3.811)$

$\qquad\quad = 1.5 \times 10^{-4}\ \text{sec}^{-1}$

Statement 5

The final products observed for many chemical reactions are formed not from a single collision among reacting particles but as the end result of a **stepwise process** involving a series of consecutive reactions. For example, an equation showing only reactants and final products and written as follows:

$$aA + bB + cC \rightarrow eE + fF$$

might actually occur in two steps:

$$\begin{array}{ll}
\text{step 1:} & aA + bB \rightarrow dD \\
\text{step 2:} & cC + dD \rightarrow eE + fF \\
\hline
\text{overall:} & aA + bB + cC \rightarrow eE + fF
\end{array}$$

D is an **intermediate** formed during the reaction, but once the reaction is complete, all D particles (molecules or ions) have been converted into E and F, the final products. The pathway defined by these steps is called the **reaction mechanism**.

The observed kinetics of such stepwise mechanisms will depend on the **relative rates** of the steps involved. The step having the highest activation energy will be the slowest step in the mechanism. This will be the **rate-determining step** of the reaction; that is, the rate of the overall reaction cannot be greater than that of this slowest step. The overall rate equation for a stepwise reaction is given by writing the rate equation for the single rate-determining step (as in Statement 1 in this Part). For example, in the following hypothetical case:

$$\begin{array}{lll}
& aA + bB \rightarrow dD & \text{(slow)} \\
& cC + dD \rightarrow eE + fF & \text{(fast)} \\
\hline
\text{overall:} & aA + bB + cC \rightarrow eE + fF &
\end{array}$$

$$\text{rate} = k\,[A]^a\,[B]^b$$

Although C is a reactant, it does not appear in the final rate expression.

If one of the steps in a mechanism is **reversible**, that is, if the product(s) of the step may regenerate the reactant(s) in the step, the rate expression is also affected. As we shall see later (Part 15, Statement 1), for a reversible process such as that shown below, we may express the relative amounts of reactants and products in that step by using an "equilibrium constant," K_e: $K_e = [D]^d/[A]^a[B]^b$. Consider this alternative to the above mechanism:

$$aA + bB \rightleftharpoons dD \quad \text{(fast and reversible)}$$
$$cC + dD \rightarrow eE + fF \quad \text{(slow)}$$

overall: $aA + bB + cC \rightarrow eE + fF$

$$\text{rate} = k\,[C]^c\,[D]^d$$

$$\text{rate} = k\,[C]^c\,(K_e[A]^a[B]^b)$$

Thus: $\text{rate} = k'\,[A]^a\,[B]^b\,[C]^c$

where $k' = kK_e$.

We see, then, that the relationship between concentrations of reactants and the rate of an overall reaction depends on its mechanism: the number of steps involved, the reversibility of a step, and which step is the one that determines the rate.

Question 13

$2H_2 + 2NO \rightarrow N_2 + 2H_2O$

One mechanism proposed for this reaction involves two steps:

$$H_2 + 2NO \rightarrow N_2O + H_2O \quad \text{(slow)}$$
$$N_2O + H_2 \rightarrow N_2 + H_2O \quad \text{(fast)}$$

a. What is the rate equation for this reaction, if this mechanism is correct?

b. What is the overall order of the reaction?

Answer 13

a. $\text{rate} = k\,[H_2]\,[NO]^2$

The first step would be rate-determining, so only the rate equation for that step need be considered.

b. third order

OEIOU: order $= a + b$

$= 1 + 2 = 3$

Question 14

The mechanism

$2NO \rightleftharpoons H_2O_2$ \quad (fast and reversible)

$$N_2O_2 + O_2 \rightarrow 2NO_2 \quad \text{(slow)}$$

has been proposed for the following reaction:

$$O_2 + 2NO \rightarrow 2NO_2$$

The rate equation for the reaction is determined experimentally to be:

$$\text{rate} = k\,[NO]^2\,[O_2]$$

Is the proposed mechanism consistent with the rate data?

Answer 14

Yes.

The second step is rate-determining. Therefore,

$$\text{rate} = k\,[N_2O_2]\,[O_2]$$
$$\text{rate} = kK_e\,[NO]^2\,[O_2]$$
$$\text{rate} = k'\,[NO]^2\,[O_2]$$

This is the same form as the experimentally determined equation.

Question 15

$$H_2O_2 + 2HI \rightarrow 2H_2O + I_2$$

The observed rate equation for the above reaction is as follows:

$$\text{rate} = k\,[H_2O_2]\,[HI]$$

Could this be the mechanism:

$$H_2O_2 + HI \rightleftharpoons H_2O + HIO \quad \text{(fast and reversible)}$$
$$HIO + HI \rightarrow H_2O + I_2 \quad \text{(slow)}$$

Answer 15

No.

The mechanism shown would lead to a different rate equation:

$$\text{rate} = k\,[HIO]\,[HI]$$
$$\text{rate} = k\,(K_e\,[H_2O_2]\,[HI]\,/\,[H_2O])\,[HI]$$
$$\text{rate} = k'\,\frac{[H_2O_2]\,[HI]^2}{[H_2O]}$$

Statement 6

Determination of the order of a given reaction from experimental data may be accomplished in several ways. One approach is to observe the effect on the reaction rate of varying the initial concentrations of reactants.

When only **one reactant** is involved, $aA \to$ products, we know that:

$$\text{rate} = k[A]_o^a$$

where a is the order to be determined and $[A]_o$ is the initial concentration of A. Therefore:

$$k = \frac{\text{rate}}{[A]_o^a}$$

By determining what value a must have in order to achieve a constant value for the rate/$[A]_o^a$ ratio, we determine the order.

When **more than one reactant** is involved, $aA + bB + \ldots \to$ products, we may carry out several experiments (performing a minimum of one more experiment than there are reactants), first keeping B_o constant in two successive experiments and solving for a:

$$\frac{\text{rate}_1}{\text{rate}_2} = \frac{k[A]_{o,1}^a [B]_{o,1}^b}{k[A]_{o,2}^a [B]_{o,1}^b} = \frac{[A]_{o,1}^a}{[A]_{o,2}^a}$$

or:

$$a = \frac{\log(\text{rate}_1/\text{rate}_2)}{\log [A]_{o,1}/[A]_{o,2}}$$

In a third experiment we keep $[A]_o$ constant and solve for b:

$$b = \frac{\log(\text{rate}_1/\text{rate}_3)}{\log [B]_{o,1}/[B]_{o,3}}$$

This procedure may be repeated in the same way for any other reactants in more complicated reactions. The important thing is to keep all concentrations constant except the one for the species whose order we are determining in a given instance.

Question 16

The data below were obtained in three experiments involving the decomposition of dimethyl ether.

a. What is the order of the reaction?

b. What is the value of k?

Experiment #	[ether]$_o$	Rate
1	0.050 M	2.2×10^{-5} mole/sec
2	0.080 M	3.4×10^{-5} mole/sec
3	0.100 M	4.3×10^{-5} mole/sec

Answer 16

a. first order

$$k = \frac{\text{rate}}{[\text{ether}]_o^a}$$

Constant value for rate/[ether]$_o^a$ exists only when $a = 1$.

b. $k = 4.3 \times 10^{-4}$ sec^{-1}

OEIOU: $k = \dfrac{\text{rate}}{[\text{ether}]_o^a}$

Question 17

The data below were obtained in experiments on the reaction between F_2 and NO_2.

a. What is the rate equation for:

$$F_2 + 2NO_2 \rightarrow 2NO_2F$$

b. What is k for the reaction?

Experiment #	[NO$_2$]$_o$	[F$_2$]$_o$	Rate
1	0.050 M	0.010 M	0.20 mole/sec
2	0.080 M	0.010 M	0.32 mole/sec
3	0.050 M	0.030 M	0.60 mole/sec

Answer 17

a. rate = $k [F_2] [NO_2]$

To find the order in NO_2:

OEIOU: $a = \dfrac{\log (\text{rate}_1/\text{rate}_2)}{\log [NO_2]_{o,1}/[NO_2]_{o,2}} = 1$

To find the order in F_2:

$$\text{OEIOU: } b = \frac{\log(\text{rate}_1/\text{rate}_3)}{\log [F_2]_{o,1}/[F_2]_{o,3}} = 1$$

b. $k = 4.0 \times 10^1$ 1 mole sec

OEIOU: $k = \text{rate}/[F_2][NO_2]$

TESTING YOUR MASTERY—Part 14 Kinetics

Question 18

Assuming the following to be one-step reactions, write rate equations for them:

a. $SO_2Cl_2 \rightarrow SO_2 + Cl_2$
b. $2NOCl \rightarrow 2NO + Cl_2$
c. $Cl_2 + CO \rightarrow COCl_2$

Answer 18

a. rate = $k [SO_2Cl_2]$
b. rate = $k [NOCl]^2$
c. rate = $k [Cl_2][CO]$

Question 19

For $2NO + Br_2 \rightarrow 2NOBr$, the rate equation is:

$$\text{rate} = k [NO]^2 [Br_2].$$

What is the order of the reaction:

a. with respect to NO?
b. with respect to Br_2?
c. overall?

Answer 19

a. second order with respect to NO
b. first order with respect to Br_2
c. third order overall

Question 20

The decomposition of N_2O_5 was studied at 27°C. In two different experiments the following data were obtained:

Experiment 1

$[N_2O_5]_{initial}$ = 0.0500 mole/L

rate = 2.00 × 10^{-6} mole/sec

Experiment 2

$[N_2O_5]_{initial}$ = 0.0750 mole/L

rate = 2.99 × 10^{-6} mole/sec

What is the overall order of the reaction

$$2N_2O_5 \rightarrow 4NO_2 + O_2$$

Answer 20

first order

Regardless of the order, rate = $k [N_2O_5]_o^a$

or

k = rate/$[N_2O_5]_o^a$

Only when a = 1 is the ratio (rate/$[N_2O_5]_o^a$) a constant value of 4 × 10^{-5}.

Question 21

For the reaction $2NOCl \rightarrow 2NO + Cl_2$ the following data for the rate of disappearance of NOCl were obtained in two successive experiments at 27°C:

Experiment #	$[NOCl]_o$	Rate
1	0.0500	4.2 × 10^{-6} mole/min
2	0.0750	9.5 × 10^{-6} mole/min
3	0.0900	1.4 × 10^{-5} mole/min

a. What is the overall order of the reaction?

b. What is the value of k?

Answer 21

a. second order

A constant value is obtained according to:

$$k = \frac{\text{rate}}{[NOCl]_o^a}$$

only when $a = 2$.

b. $k = 1.7 \times 10^{-3}$ L mole^{-2} min^{-1}

OEIOU: $k = \dfrac{\text{rate}}{[NOCl]_o^a}$

Question 22

Hydrogen peroxide decomposes according to this equation:

$$2H_2O_2 \rightarrow 2H_2O + 2O_2$$

If $k = 4.7 \times 10^{-4}$ L/mole sec, and the initial quantity is 0.100 mole/L of H_2O_2, what is the half-life of the reaction?

Answer 22

2.1×10^4 sec

OEIOU: $t_{\frac{1}{2}} = \dfrac{1}{k[H_2O_2]_o}$

Question 23

The high-temperature (873°K) decomposition of CH_2N_2 occurs with a rate constant of 9.2×10^{-4} sec^{-1}. If one starts with 0.100 mole/L of CH_2N_2, how much time must elapse until the concentration drops to half this value?

Answer 23

7.5×10^2 sec

OEIOU: $t_{\frac{1}{2}} = \dfrac{0.693}{k}$

Question 24

Ethyl chloride, C_2H_5Cl, decomposes according to the following equation:

$$C_2H_5Cl \rightarrow HCl + C_2H_4$$

Activation energy for the process is 59 kcal and the Arrhenius factor, A, is 1.6×10^{14} sec^{-1}. Predict k for the decomposition at 600°C.

Answer 24

2.9×10^{-1} sec^{-1}

OEIOU: $k = $ antilog $\log A - \dfrac{E_a}{2.303\, RT}$

Question 25

If the temperature were dropped from 600°C to 500°C in Question 24, what would be

a. the value of k

b. the rate of decomposition of C_2H_5Cl, if the initial concentration were 0.200 mole/L?

Answer 25

a. 3.6×10^{-3} sec^{-1}

OEIOU: $k_2 = $ antilog $\left[\dfrac{E_a(T_2 - T_1)}{2.303\, RT_2 T_1} + \log k_1 \right]$

b. 7.2×10^{-4} mole/sec

OEIOU: rate$_2$ = $k_2 [C_2H_5Cl]$

Question 26

Show the rate expression for this reaction: $O_2 + 2NO \rightarrow 2NO_2$ if the mechanism is:

$O_2 + NO \rightleftharpoons NO_3$ (fast and reversible)

$NO_3 + NO \rightarrow 2NO_2$ (slow)

Answer 26

$$\text{rate} = k[NO]^2[O_2]$$

(Compare with Question 14.)

Question 27

The rate equation for the reaction between Cl_2 and CO to give $COCl_2$ is:

$$\text{rate} = k[CO][Cl_2]^{\frac{3}{2}}$$

Is the following mechanism feasible?

$$Cl_2 \rightleftharpoons 2Cl \quad \text{(fast and reversible)}$$
$$Cl + CO \rightleftharpoons COCl \quad \text{(fast and reversible)}$$
$$COCl + Cl_2 \rightarrow COCl_2 + Cl \quad \text{(slow)}$$

Answer 27

Yes.

Derivation of the rate equation:

$$\text{rate} = k[COCl][Cl_2]$$

$$\text{rate} = kK_{e,2}[CO][Cl][Cl_2]$$

$$\text{rate} = kK_{e,2}[CO]\,K_{e,1}^{\frac{1}{2}}[Cl]^{\frac{1}{2}}[Cl][Cl_2]$$

thus:

$$\text{rate} = k'[CO][Cl]^{\frac{3}{2}}$$

PART 15
EQUILIBRIA INVOLVING GASES

Statement 1

It is common practice to write a chemical equation with a single arrow pointing from "reactants" on the left to "products" on the right. In theory, however, we could join the two sides of the equation by **double arrows**, one pointing to the left, the other to the right, indicating that the reaction is reversible, and that the products might also react with one another to regenerate reactants.

$$\text{reactants} \quad aA + bB \rightleftharpoons cC + dD \quad \text{products}$$

(where a, b, c, and d are coefficients in the balanced equation)

If both the forward and reverse processes take place, the forward process slows down as A and B are used up, and the reverse process speeds up as C and D are generated; when the forward and reverse rates become identical, **equilibrium** has been attained.

The position of a system at equilibrium is measured by the **equilibrium constant**,* K_e, given by:

Law of Chemical Equilibrium $$K_e = \frac{[C]^c[D]^d}{[A]^a[B]^b}$$

where the terms in brackets are **molar concentrations** of reactants and products. Concentrations remain unchanged once equilibrium is attained, provided that nothing happens (e.g., change in temperature) to disturb the equilibrium.

It should be noted that an equation may be "balanced" by numerous choices of a, b, c, d, and so forth. For example, all of the following are balanced equations:

$$C_2H_6 + \frac{7}{2}O_2 \rightleftharpoons 2CO_2 + 3H_2O$$

$$2C_2H_6 + 7O_2 \rightleftharpoons 4CO_2 + 6H_2O$$

$$4C_2H_6 + 14O_2 \rightleftharpoons 8CO_2 + 12H_2O$$

The value of K_e is determined by which coefficients we choose in balancing the equation of interest. Normally, an equation is balanced by using as coefficients the smallest possible whole numbers.

---

*The equilibrium constant is symbolized by K_c in some texts.

PART 15—EQUILIBRIA INVOLVING GASES

Question 1

Phosphorous pentachloride gas decomposes to some extent at 150°C to give phosphorous trichloride and chlorine. Write the equilibrium expression for this process according to the law of chemical equilibrium. (Both products are gases.)

Answer 1

$$K_e = \frac{[PCl_3][Cl_2]}{[PCl_5]}$$

Balanced equation: $PCl_5(g) \rightleftharpoons PCl_3(g) + Cl_2(g)$

Question 2

A certain quantity of PCl_5 gas is heated to 150°C. If at equilibrium the molar concentrations of PCl_5, PCl_3, and Cl_2 gases are 0.0080 M, 0.020 M, and 0.020 M, respectively, what is K_e for the system:

$$PCl_5(g) \rightleftharpoons PCl_3(g) + Cl_2(g) \ ?$$

Answer 2

0.050

OEIOU: $K_e = \dfrac{[PCl_3][Cl_2]}{[PCl_5]}$

$= \dfrac{(0.020)(0.020)}{(0.0080)}$

$= 0.050$

Question 3

A certain quantity of Cl_2 and PCl_3 are heated to 150°C. If molar concentrations of PCl_3 and Cl_2 are each 0.010 M at equilibrium, what is the concentration of PCl_5?
(Refer back to Answer 2 if necessary.)

Answer 3

0.0020 mole/L

OEIOU: $[PCl_5] = \dfrac{[PCl_3][Cl_2]}{K_e}$

Question 4

Write the equilibrium expression for the formation of ammonia from nitrogen and hydrogen.

Answer 4

$$K_e = \frac{[NH_3]^2}{[N_2][H_2]^3}$$

Balanced equation: $N_2(g) + 3H_2(g) \rightleftharpoons 2NH_3(g)$

Question 5

K_e for the process $2NO_2(g) \rightleftharpoons N_2O_4(g)$ is 2.2×10^2 at 25°C. If the equilibrium concentration of N_2O_4 in a certain container is 3.5×10^{-2} mole/L, what is the concentration of NO_2?

Answer 5

1.3×10^{-2} mole/L

OEIOU: $[NO_2] = \sqrt{\dfrac{[N_2O_4]}{K_e}}$

Question 6

If K_e for the process $2NO_2(g) \rightleftharpoons N_2O_4(g)$ is 2.2×10^2 at 25°C, what is K_e for the process $N_2O_4(g) \rightleftharpoons 2NO_2(g)$?

Answer 6

4.5×10^{-3}

$$K_e' = \frac{[NO_2]^2}{[N_2O_4]} = \frac{1}{[N_2O_4]/[NO_2]^2}$$

OEIOU: $K_e' = \dfrac{1}{K_e}$

$= 1/(2.2 \times 10^2)$

$= 4.5 \times 10^{-3}$

Statement 2

We may use a knowledge of the value of K_e for a process to predict the **direction of reaction** when given initial concentrations of reactants and products are mixed. From these given concentrations, a **concentration quotient**, Q_c, may be calculated for the process

$$aA + bB \rightleftharpoons cC + dD$$

$$Q_c = \frac{[C]_i^c \, [D]_i^d}{[A]_i^a \, [B]_i^b}$$

where the i subscripts refer to **initial** (as opposed to equilibrium) concentrations. Then:

if $Q_c < K_e$, the reaction goes to the **right**;

if $Q_c > K_e$, the reaction goes to the **left**; and

if $Q_c = K_e$, the reaction is at **equilibrium**.

Question 7

If 0.10 moles each of PCl_5, PCl_3, and Cl_2 gases are mixed in a one-liter vessel at 150°C, what is Q_c for the process:

$$PCl_5(g) \rightleftharpoons PCl_3(g) + Cl_2(g) \text{ ?}$$

Answer 7

0.10

OEIOU: $Q_c = \dfrac{[PCl_3]_i \, [Cl_2]}{[PCl_5]_i}$

$= \dfrac{(0.10 \text{ M}) (0.10 \text{ M})}{(0.10 \text{ M})}$

$= 0.10$

Question 8

When 0.10 mole each of PCl_5, PCl_3, and Cl_2 gases are mixed in a one-liter vessel at 150°C, would some of the PCl_5 decompose, or would more PCl_5 be formed? ($K_e = 5.0 \times 10^{-2}$.)

Answer 8

More PCl$_5$ would be *formed*.

We see that $Q_c > K_e$ for the decomposition process:

$$Q_c = 0.10 > K_e = 0.050$$

Question 9

If 0.10 mole each of NO$_2$ and N$_2$O$_4$ are mixed in a two-liter vessel at 25°C, would the concentration of N$_2$O$_4$ increase or decrease? ($K_e = 2.2 \times 10^2$ for the formation of N$_2$O$_4$ from NO$_2$.)

Answer 9

Increase.

$$Q_c = \frac{[N_2O_4]_i}{[NO_2]_i^2}$$

$$= \frac{(\#\,\text{moles}\,N_2O_4/\#\,L)}{(\#\,\text{moles}\,NO_2/\#\,L)^2}$$

$$= \frac{(0.10\,\text{mole}/2\,L)}{(0.10\,\text{mole}/2\,L)^2}$$

$$= 20$$

Thus: $Q_c < K_e$

Statement 3

The **extent of reaction** is indicated to an approximate extent by the size of the equilibrium constant. In general:

if K_e is **very large**, the reaction essentially goes to **completion**;

if K_e is **very small**, there is essentially **no reaction**; and

if K_e is **intermediate** in value, there is **partial reaction**, with significant quantities of both reactants and products **mixed** at equilibrium.

Question 10

At 25°C, K_e for the process $3O_2(g) \rightleftharpoons 2O_3(g)$ is 1.0×10^{-55}. Does the transformation of oxygen to ozone essentially go to completion at 25°C?

Answer 10

No.

1.0×10^{-55} is a *very small* number; therefore there is virtually no reaction in the forward direction.

Question 11

At 400°C, K_e for the process $2SO_2(g) + O_2(g) \rightleftharpoons 2SO_3(g)$ is 7.2×10^{-1}. Can SO_2 be converted essentially 100% into SO_3 at this temperature?

Answer 11

No.

7.2×10^{-1} is neither very large nor very small, so a significant quantity of SO_2 will remain unreacted.

Question 12

At 127°C, K_e for the process $2H_2(g) + O_2(g) \rightleftharpoons 2H_2O(g)$ is 3.0×10^{58}. Would the reaction between hydrogen and oxygen be a good method for forming water at 127°C?

Answer 12

Yes.

3.0×10^{58} is a very large number; therefore there is essentially 100% conversion of H_2 and O_2 to H_2O.

Statement 4

More exact calculations of **equilibrium concentrations** of reactants and products may be carried out using K_e values, allowing us to predict exactly how much reactant will remain or how much product will be formed at equilibrium.

For example, for this type of equilibrium:

$$A_2(g) \rightleftharpoons 2A(g)$$

for each mole of A_2 dissociating, **two** moles of A_2 will form. Hence, if *a* represents the number of moles of A_2 originally present and n the number of moles of A_2 which dissociate, the equilibrium concentration of A will be:

$$[A]_e = 2n/v$$

and for A_2 at equilibrium: $\quad [A_2]_e = (a-n)/v$

where v is the volume of the reaction vessel. Putting these into the equilibrium expression, one may solve for n, and from this find $[A]_e$ and $[A_2]_e$. A similar approach may be used with other equilibria.

Question 13

If 1.0 mole each of H_2 and I_2 are mixed in a ten-liter vessel at 520°C, how many moles each of H_2 and I_2 will have reacted at equilibrium? (K_e for the process H_2 (g) + I_2 (g) $\rightleftharpoons$ 2HI (g) is 62.5 at 520°C.)

Answer 13

0.80 mole H_2 and 0.80 mole I_2

$$K_e = [HI]^2/[H_2][I_2]$$

OEIOU: $K_e = (2n)^2/(a-n)(b-n)$

where: a = moles H_2, initial
b = moles I_2, initial
n = moles H_2, reacting
  = moles I_2, reacting

Thus: $n^2(4-K_e) + K_e(a+b)n - K_e ab = 0$

Solving by quadratic formula: $n = 0.80$

Note: One gets two answers, 0.80 and 1.33, but the second is obviously impossible.

Question 14

What are the equilibrium concentrations at 520°C of H_2, I_2, and HI in Question 13 above?

Answer 14

$$[H_2] = [I_2] = 0.020 \text{ M}$$

$[HI] = 0.16 \text{ M}$

OEIOU: $[H_2] = (a-n)/v$

OEIOU: $[I_2] = (b-n)/v$

OEIOU: $[HI] = 2n/v$

Question 15

If $K_e = 4.5 \times 10^{-3}$ at 25°C for $N_2O_4(g) \rightleftharpoons 2NO_2(g)$, what are the equilibrium concentrations of N_2O_4 and NO_2 at 25°C if the confining vessel has a volume of ten liters and if 0.50 moles of N_2O_4 are initially present?

Answer 15

$[N_2O_4] = 0.043 \text{ M}$ $[NO_2] = 0.014 \text{ M}$

$$K_e = \frac{(2n/v)^2}{(a-n)/v}$$

$$= \frac{4n^2}{(a-n)v}$$

where a = moles N_2O_4, initial

n = moles N_2O_4, reacting

Thus: $4n^2 + K_e vn - K_e va = 0$

Solving: $n = 0.070 \text{ M}$

(The other value obtained in solving the quadratic, $n = -0.080 \text{ M}$, is impossible; a concentration cannot be negative!)

Then find the concentrations of N_2O_4 and NO_2 from:

$[N_2O_4] = (a-n)/v$

$[NO_2] = 2n/v$

Statement 5

Although for a given temperature K_e is a constant, individual concentrations may be changed, provided that all other concentrations change also, so that K_e remains "satisfied."

PART 15–EQUILIBRIA INVOLVING GASES

A **shift in equilibrium** will result, then, from **changing the concentration** of a reactant or product involved in the equilibrium. Specifically:

if the concentration of a species is **increased**, equilibrium will shift to **consume** part of that species;

if the concentration of a species is **decreased**, equilibrium will shift to **build up** its concentration again.

Question 16

If the system $2ICl\,(g) \rightleftharpoons I_2\,(g) + Cl_2\,(g)$ is at equilibrium, and the concentration of I_2 is increased, in which direction will the equilibrium shift?

Answer 16

Left.

A concentration is increased on the "product" side of the equation, so equilibrium will shift in the opposite direction to consume some of the added I_2.

Question 17

If the system $N_2\,(g) + 3H_2\,(g) \rightleftharpoons 2NH_3\,(g)$ is at equilibrium, and additional N_2 is suddenly introduced, in which direction will the equilibrium shift?

Answer 17

Right.

The shift consumes some of the added N_2.

Question 18

If the system $2CO\,(g) + O_2\,(g) \rightleftharpoons 2CO_2\,(g)$ is at equilibrium, and some of the CO_2 is removed, in which direction will equilibrium shift?

Answer 18

Right.

Removing CO_2 decreases the concentration of CO_2, the product. Therefore, the equilibrium will shift to that side of the arrows, producing more CO_2.

Question 19

A 1.00-liter vessel contains 0.240 mole HI, 0.030 mole H_2, and 0.030 mole I_2 in equilibrium at 520°C. K_e is 62.5 for the system: I_2 (g) + H_2 (g) ⇌ 2HI (g) at 520°C. If the concentration of HI is suddenly doubled, in which direction will equilibrium shift?

Answer 19

Left.

Question 20

How many moles of HI will react upon doubling the concentration of HI at 520°C in Question 19?

Answer 20

0.049 mole

$$K_e = \frac{(c-n)^2}{(a+\frac{1}{2}n)(b+\frac{1}{2}n)}$$

where a = moles H_2, initial

b = moles I_2, initial

c = moles $HI_{initial}$ (0.48 mole)

n = moles HI reacting

Because in this problem, $a = b$:

$$K_e = \frac{(c-n)^2}{(a+\frac{1}{2}n)^2}$$

$$K_e = \frac{(c-n)^2}{(a+\frac{1}{2}n)^2}$$

OEIOU: $n(1 + \frac{1}{2}\sqrt{K_e}) + (a\sqrt{K_e} - C) = 0$

Solve for n: n = 0.049 moles

Question 21

What will be the equilibrium concentrations of HI, I_2, and H_2 after doubling the concentration of HI in Question 19?

Answer 21

[HI] = 0.43 M

[H$_2$] = 0.054 M

[I$_2$] = 0.054

OEIOU: [HI] = $(c - n)/v$

OEIOU: [H$_2$] = $(a + \frac{1}{2}n)/v$

OEIOU: [I$_2$] = $(b + \frac{1}{2}n)/v$

Statement 6

A shift in equilibrium may also result from **changing the volume** to which gaseous reactants and products are confined. In general, when the number of **moles of gaseous reactants and products differ**:

if the **volume increases**, equilibrium will shift to **increase** the number of moles of gas;
if the **volume decreases**, equilibrium will shift to **decrease** the number of moles of gas.

Question 22

If the system N$_2$ (g) + 3H$_2$ (g) ⇌ 2NH$_3$ (g) is in equilibrium at 300°C and the volume of the container is suddenly changed from 10 liters to 30 liters, in which direction will the equilibrium shift?

Answer 22

Left.

4 moles of gases on left, 2 moles of gases on right. Increased volume favors increase in total moles of gases.

Question 23

If the volume of the system 3O$_2$ (g) ⇌ 2O$_3$ (g) is suddenly changed from 20.0 liters to 15.0 liters, will more ozone be formed?

Answer 23

Yes.

3 moles of gases on left, 2 moles of gases on right. Decreased volume favors shift toward side occupying less space.

Question 24

If the system $H_2(g) + I_2(g) \rightleftharpoons 2HI(g)$ is in equilibrium at room temperature in a one-liter vessel, and if the volume is suddenly doubled, will more HI be formed, or will some HI react?

Answer 24

Neither.

2 moles of gases on left, 2 moles of gases on right. Neither side favored.

Question 25

If 0.88 mole N_2O_4 and 0.020 mole NO_2 are in equilibrium at 25°C in a one-liter vessel, and if the volume suddenly changes from 1.0 to 2.0 liters, will more N_2O_4 decompose?

Answer 25

Yes.

$2NO_2(g)$ occupies more space than $N_2O_4(g)$.

Question 26

If K_e for the system $N_2O_4(g) \rightleftharpoons 2NO_2$ is 4.5×10^{-3} at 25°C, how much N_2O_4 will remain after equilibrium is re-established in Question 25?

Answer 26

0.85 mole

$$K_e = \frac{(b + 2n)^2}{(a - n)v}$$

where a = moles N_2O_4, initial

b = moles NO_2, initial

n = moles N_2O_4, reacting

Thus: $4n^2 + n(4b + K_e v) + (b^2 - K_e va) = 0$

Solving: $n = 0.034$ mole

Then: moles N_2O_4, final = $a - n$

= 0.88 mole − 0.034 mole

= 0.85 mole

Statement 7

When an equilibrium involves a solid or liquid as well as a gas, the equilibrium expression is written in the usual way, except that the concentration term for the solid or liquid does not appear in the expression.

Example

$A(s) + B(g) \rightleftharpoons C(g) + D(g)$ a heterogeneous equilibrium

$$K_e = \frac{[C][D]}{[B]}$$

Question 27

Write the K_e expression for the process

$$H_2O(l) + CO(g) \rightleftharpoons CO_2(g) + H_2(g)$$

Answer 27

$$K_e = \frac{[H_2][CO_2]}{[CO]}$$

Question 28

Write the K_e expression for the process

$$MgCO_3(s) \rightleftharpoons CO_2(g) + MgO(s)$$

Answer 28

$$K_e = [CO_2]$$

Question 29

Acetylene gas, C_2H_2, is produced by the process:

$$CaC_2(s) + 2H_2O(l) \rightleftharpoons Ca(OH)_2(s) + C_2H_2(g)$$

If equilibrium is established and more water is added, will the equilibrium shift?

Answer 29

No.

$K_e = [C_2H_2]$, so unless K_e is changed, the concentration of acetylene gas cannot change. K_e is unaffected by the amount of liquid water present.

Statement 8

An alternative expression of the position of a gaseous system at equilibrium,

$$aA(g) + bB(g) \rightleftharpoons cC(g) + dD(g)$$

involves the use of partial pressures (in atm) rather than molar concentrations:

$$K_p = \frac{p_C^c \, p_D^d}{p_A^a \, p_B^b}$$

The symbol K_p is used rather than K_e (or K_c, as many people designate it) because the two constants will often have different values. K_p and K_e are related by the equation

$$K_p = K_e(RT)^{\Delta n_g}$$

where R is the gas constant (0.0821 L·atm/mole°K) and Δn_g is the change in the number of moles of gas:

$$\Delta n_g = (c + d) - (a + b)$$

Question 30

If $K_e = 2.2 \times 10^2$ at 25°C for the process

$$2NO_2(g) \rightleftharpoons N_2O_4(g)$$

what is the value of K_p?

Answer 30

9.0

$$K_p = K_e(RT)^{\Delta n_g}$$

OEIOU: $K_p = K_e (RT)^{c-a}$

$= (2.2 \times 10^2)(0.0821 \times 298)^{1-2}$

$= \dfrac{2.2 \times 10^2}{0.0821 \times 298}$

$= 9.0$

Question 31

K_p at 25°C for the process

$O_2(g) + 2SO_2(g) \rightleftharpoons 2SO_3(g)$ is 1.2×10^{22}

What is the value of K_e?

Answer 31

2.9×10^{23}

$$K_p = K_e(RT)^{\Delta n_g}$$

OEIOU: $K_e = K_p/(RT)^{c-a-b}$

$= \dfrac{1.2 \times 10^{22}}{(8.21 \times 10^{-2} \times 2.98 \times 10^2)^{2-1-2}}$

$= 1.2 \times 10^{22} \times 8.21 \times 10^{-2} \times 2.98 \times 10^2$

$= 2.9 \times 10^{23}$

Question 32

K_e is 1.0×10^{44} at 25°C for the process $O_2(g) + CuS(s) \rightleftharpoons Cu(s) + SO_2(g)$. Calculate K_p.

Answer 32

1.0×10^{44}

OEIOU: $K_p = K_e (RT)^{c+d-a-b}$

$= (1.0 \times 10^{44})(0.0821 \times 298)^{0+1-1-0}$

$= 1.0 \times 10^{44}$

Statement 9

K_p is related to the standard free energy of reaction by the expression

$$\Delta G^\circ_{rx} = -RT \ln K_p$$

(where ΔG° is the free energy change when reactants and products are at 1 atmosphere) so that it becomes possible to calculate equilibrium constants from thermochemical data alone. $R = 1.99$ cal/mole °K in this equation.

Question 33

ΔG°_f is -22.8 kcal/mole for the formation of HCl gas at 25°C. What is K_p for the reaction?

$$H_2(g) + Cl_2(g) \rightleftharpoons 2HCl(g) \;?$$

Answer 33

2.46×10^{33}

Two moles of HCl are being formed, so:

$$2\Delta G^\circ_f = \Delta G^\circ_{rx} = -RT \ln K_p$$

$$2\Delta G^\circ_f = -2.303\, RT \log K_p$$

OEIOU: $K_p = \text{antilog} \left[\dfrac{2\Delta G^\circ_f}{-2.303\, RT} \right]$

$ = \text{antilog} \left[\dfrac{2(-22{,}800 \text{ cal})}{-2.303\, (1.99 \text{ cal/mole}°K)(298°K)} \right]$

$ = \text{antilog} (33.389)$

$ = 2.46 \times 10^{33}$

Question 34

$\Delta G^\circ_{f,NH_3}$ is -3.98 kcal/mole. What is K_e for this reaction at 25°C:

$$N_2(g) + 3H_2(g) \rightleftharpoons 2NH_3(g)$$

Answer 34

4.03×10^8

$$2\Delta G^\circ_f = \Delta G^\circ_{rx} = -RT \ln K_p$$

$$2\Delta G_f^\circ = -2.303\,RT\log\left[K_e(RT)^{\Delta n_g}\right]$$

$$K_e(RT)^{\Delta n_g} = \text{antilog}\left[\frac{2\Delta G_f^\circ}{-2.303\,RT}\right]$$

OEIOU: $$K_e = \frac{\text{antilog}\left[\frac{2\Delta G_f^\circ}{-2.303\,RT}\right]}{(RT)^{\Delta n_g}}$$

$$= \frac{\text{antilog}\,(5.828)}{(24.47)^{-2}}$$

$$= 4.03 \times 10^8$$

TESTING YOUR MASTERY—Part 15
Equilibria Involving Gases

Question 35

If SO_3 gas decomposes on heating into SO_2 and O_2,

a. write the balanced equation for the process.

b. write the equilibrium expression for the process according to the law of chemical equilibrium.

Answer 35

a. $2SO_3(g) \rightleftarrows 2SO_2(g) + O_2(g)$

b. $K_e = \dfrac{[SO_2]^2\,[O_2]}{[SO_3]^2}$

Question 36

A sample of SO_3 gas is heated to 600°C. At equilibrium the molar concentrations of SO_3, SO_2, and O_2 are 0.00100 M, 0.0026 M, and 0.0013 M, respectively. What is the value of K_e for this SO_3 decomposition?

Answer 36

8.7×10^{-3}

OEIOU: $K_e = \dfrac{[SO_2]^2\,[O_2]}{[SO_3]^2}$

Question 37

For the process $COCl_2(g) \rightleftarrows CO(g) + Cl_2(g)$, $K_e = 2.6 \times 10^{-10}$ at $100°C$. A mixture of $COCl_2$ and Cl_2 are heated to $100°$. At equilibrium their molar concentrations are 0.028 M and 0.00013 M, respectively. What is the CO concentration at equilibrium?

Answer 37

5.6×10^{-8} M

OEIOU: $[CO] = \dfrac{K_e [COCl_2]}{[Cl_2]}$

Question 38

If K_e for the decomposition of phosgene:

$$COCl_2(g) \rightleftarrows CO(g) + Cl_2(g)$$

is 2.6×10^{-10} at $100°C$,

a. what is K_e' for the following process:

$$CO(g) + Cl_2(g) \rightleftarrows COCl_2(g)$$

b. Does the *formation* of phosgene essentially go to completion?

Answer 38

a. 3.8×10^9

OEIOU: $K_e' = \dfrac{1}{K_e}$

b. Yes. The very large value of K_e' for the formation of phosgene indicates a strong tendency of CO and Cl_2 to react to form $COCl_2$.

Question 39

If 0.100 mole/L of $COCl_2$ is heated to $100°C$, what is the equilibrium concentration of:

a. CO?

b. Cl_2?

c. $COCl_2$?

(K_e for the decomposition of $COCl_2$ at 100°C is 2.6×10^{-10}.)

Answer 39

a. 5.1×10^{-6} M = $[CO]_e$

b. 5.1×10^{-6} M = $[Cl_2]_e$

OEIOU: $K_e = \dfrac{[CO]_e^2}{[COCl_2]_i - [CO]_e}$

Solve for $[CO]_e$. Note that since one Cl_2 is formed each time one CO is formed, $[Cl_2]_e = [CO]_e$.

c. $[COCl_2] \approx 0.100$ mole/L

Because K_e is so small, $[CO]_e$ is very small also, and very little of the $COCl_2$ decomposes.

Question 40

Methyl alcohol, CH_3OH, may be prepared by heating CO and H_2 gases:

$$CO(g) + 2H_2(g) \rightleftarrows CH_3OH(g)$$

If K_e for the reaction at 700°C is 7.0, and if 0.0100 mole each of CO, H_2, and CH_3OH are mixed in a one-liter vessel,

a. what would be Q_c?

b. would more CH_3OH form?

Answer 40

a. $Q_c = 10{,}000$

OEIOU: $Q_c = \dfrac{[CH_3OH]_i}{[CO]_i \, [H_2]_i^2}$

b. No. Because $Q_c > K_e$, the equilibrium would shift to the *left*, forming more CO and H_2.

Question 41

K_e is 0.16 for this reaction at 25°C:

$$2NOBr(g) \rightleftarrows 2NO(g) + Br_2(g)$$

If 0.100 mole of NOBr, 0.0200 mole of NO, and 0.0400 mole of Br_2 are mixed in a two-liter vessel at 25°C, would more NO and Br_2 form?

Answer 41

Yes.

Since $Q_c < K_e$, shift to the *right*.

$$\text{OEIOU:} \quad Q_c = \frac{(n_{NO}/v)^2 \, (n_{Br_2}/v)}{(n_{NOBr}/v)^2}$$

$$\text{or:} \quad Q_c = \frac{n_{NO}^2 \times n_{Br_2}}{n_{NOBr}^2 \times v}$$

Question 42

If 0.15 mole Cl_2 and 0.30 mole PCl_3 are mixed in a 5.00-liter vessel at 150°C, at equilibrium,

a. how many moles of PCl_3 would have reacted?
b. how many moles of Cl_2 would have reacted?
c. how many moles of PCl_5 would have formed?

K_e is 0.050 at 150° for:

$$PCl_5(g) \rightleftarrows PCl_3(g) + Cl_2(g)$$

Answer 42

a. 0.070 moles PCl_3 reacted
b. 0.070 moles Cl_2 reacted
c. 0.070 moles PCl_5 formed

$$\text{OEIOU:} \quad K_e = \frac{(a-n)(b-n)}{nv}$$

where: a = moles $Cl_{2,\,initial}$
b = moles $PCl_{3,\,initial}$
n = moles $Cl_{2,\,reacting}$
    = moles $PCl_{3,\,reacting}$
    = moles $PCl_{5,\,formed}$

Question 43

Referring back to Question 42, when 0.15 mole Cl_2 and 0.30 mole PCl_3 are mixed in a 5.00-liter vessel at 150°C, what would be the equilibrium concentration of:

a. PCl_3?

b. Cl_2?

c. PCl_5?

Answer 43

a. 0.046 M

OEIOU: $[PCl_3]_e = \dfrac{b - n}{V}$

b. 0.016 M

OEIOU: $[Cl_2]_e = \dfrac{a - n}{V}$

c. 0.014 M

OEIOU: $[PCl_5]_e = \dfrac{n}{V}$

Question 44

Suppose 0.070 mole of PCl_3, 0.080 mole of Cl_2, and 0.23 mole of PCl_5 are in equilibrium in a 5.00-liter vessel at 150°C. If the concentration of PCl_5 is suddenly doubled by introducing more PCl_5,

a. in which direction will the equilibrium shift in Question 42?

b. how many moles of PCl_5 will react or be formed?

c. how many moles of PCl_3 will react or be formed?

d. how many moles of Cl_2 will react or be formed?

Answer 44

a. Shift to *right*, to form more PCl_3 and Cl_2.

b. 0.18 moles of PCl_5 will react.

OEIOU: $K_e = \dfrac{(a + n)(b + n)}{(c - n) V}$

where: a = moles Cl_2, initial

b = moles PCl_3, initial

c = moles PCl_5, initial (0.46 mole)

n = moles PCl_5, reacting

Solve for n.

c. 0.18 mole PCl_3 formed.

d. 0.18 mole Cl_2 formed.

Question 45

At 25°C the K_e is 0.0045 for the process:

$$N_2O_4(g) \rightleftarrows 2NO_2(g)$$

If 0.12 mole NO_2 and 0.32 mole N_2O_4 are in equilibrium in a ten-liter vessel and the concentration of NO_2 is increased by 50%,

a. in which direction will the equilibrium shift?

b. how many moles of NO_2 will react or be formed?

c. how many moles of N_2O_4 will react or be formed?

Answer 45

a. Shift to *left*, to form more N_2O_4.

b. 0.055 mole NO_2 reacts.

OEIOU: $K_e = \dfrac{(a-n)^2}{\left(b + \dfrac{1}{2}n\right)v}$

where: a = moles NO_2, initial (0.18 mole)

b = moles N_2O_4, initial

n = moles NO_2, reacting

c. 0.028 mole N_2O_4 formed.

OEIOU: # moles N_2O_4 formed = $\dfrac{1}{2}$ # moles NO_2 reacting

---

Question 46

K_e is 0.050 at 150°C for:

$$PCl_5(g) \rightleftarrows PCl_3(g) + Cl_2(g).$$

If the equilibrium concentrations of PCl_5, PCl_3, and Cl_2 are 0.0047 M, 0.030 M, and 0.0078 M, respectively, and the volume of the vessel containing the gases is changed from 1.00 to 4.00 liters,

a. in which direction will the equilibrium shift?

b. how many moles of PCl_5 will react or form?

c. how many moles each of PCl_3 and Cl_2 will react or form?

d. what will be the equilibrium concentration of PCl_5?

Answer 46

a. Shift to *right* since there are more moles of gas on the right than the left and the volume is *increasing*.

b. 0.0030 mole PCl_5 reacts.

OEIOU: $K_e = \dfrac{(a + n)(b + n)}{(c - n)v}$

where: a = moles Cl_2, initial

b = moles PCl_3, initial

c = moles PCl_5, initial

n = moles PCl_5, reacting

= moles PCl_3 formed

= moles Cl_2 formed

Solve for n.

c. 0.0030 mole each of PCl_3 and Cl_2 formed.

d. 4.2×10^{-4} M

OEIOU: $[PCl_5] = \dfrac{c - n}{v}$

Question 47

Which of the following processes will be shifted to the right by *decreasing* the volume of the vessel containing the reaction mixture?

a. $NO_2(g) + NO_3(g) \rightleftarrows N_2O_5(g)$

b. $2ClO(g) \rightleftarrows Cl_2(g) + O_2(g)$

c. $2NO(g) + Cl_2(g) \rightleftarrows 2NOCl(g)$
d. $CaO(s) + CO_2(g) \rightleftarrows CaCO_3(s)$

Answer 47

a, c, and d

Each of these occurs with a decrease in required volume as written, thus is favored by decreasing the available volume.

Question 48

For the following equilibria, express K_e in terms of molar concentrations:

a. $CaO(s) + CO_2(g) \rightleftarrows CaCO_3(s)$
b. $Fe_3O_4(s) + 4H_2(g) \rightleftarrows 3Fe(s) + 4H_2O(g)$
c. $CO_2(g) + H_2(g) \rightleftarrows CO(g) + H_2O(g)$

Answer 48

a. $K_e = \dfrac{1}{[CO_2]}$

b. $K_e = \dfrac{[H_2O]^4}{[H_2]^4}$

c. $K_e = \dfrac{[CO][H_2O]}{[CO_2][H_2]}$

The first two equilibria are *heterogeneous* equilibria, with the solids not appearing in the equilibrium expressions.

Question 49

The equilibrium pressures of SO_3, SO_2, and O_2 in a certain mixture at 25°C are 1.2 atm, 0.84 atm, and 1.7×10^{-22} atm, respectively. What is K_p for this process?

$$2SO_3(g) \rightleftarrows 2SO_2(g) + O_2(g)$$

Answer 49

8.3×10^{-23}

OEIOU: $K_p = \dfrac{P_{SO_2}^2 \times P_{O_2}}{P_{SO_3}^2}$

Question 50

Calculate K_p for these processes:

a. $COCl_2(g) \rightleftarrows CO(g) + Cl_2(g)$

$K_e = 2.6 \times 10^{-10}$ at $100°C$

b. $CO(g) + 2H_2(g) \rightleftarrows CH_3OH(g)$

$K_e = 7.0$ at $700°C$

c. $2NOBr(g) \rightleftarrows 2NO(g) + Br_2(g)$

$K_e = 0.16$ at $25°C$

d. $NH_4SH(s) \rightleftarrows NH_3(g) + H_2O(g)$

$K_e = 8.7 \times 10^{-5}$ at $20°C$

Answer 50

a. 8.0×10^{-9}

OEIOU: $K_p = K_e(RT)^{c+d-a}$

b. 1.1×10^{-3}

OEIOU: $K_p = K_e(RT)^{c-(a+b)}$

c. 4.0

OEIOU: $K_p = K_e(RT)^{c+d-a}$

d. 5.0×10^{-2}

OEIOU: $K_p = K_e(RT)^{c+d}$

Question 51

$K_p = 7.0$ for this process at $1000°C$:

$$C(s) + CO_2(g) \rightleftarrows 2CO(g)$$

What is K_e?

Answer 51

6.7 × 10^{-2}

OEIOU: $K_e = \dfrac{K_p}{(RT)^{c-b}}$

Question 52

Calculate ΔG°_{rx} for these reactions at 25°C:

a. $2NOBr(g) \rightleftarrows 2NO(g) + Br_2(g)$

   $K_p = 4.0$

b. $2SO_3(g) \rightleftarrows 2SO_3(g) + O_2(g)$

   $K_p = 8.3 \times 10^{-23}$

Answer 52

a. −822 cal

OEIOU: $\Delta^\circ_{rx} = RT \ln K_p$

b. +30 kcal

Question 53

$\Delta G^\circ_{f,N_2O}$ is +25 kcal/mole. What is K_p for this reaction at 25°C?

$$2N_2(g) + O_2(g) \rightleftarrows 2N_2O(g)$$

Answer 53

2.4 × 10^{-37}

OEIOU: $K_p = \text{antilog}\left[\dfrac{-2\Delta G^\circ_f}{2.303\, RT}\right]$

$ = \text{antilog}(-36.611)$

$ = 2.4 \times 10^{-37}$

Question 54

$\Delta G°$ is +26 kcal/mole at 25°C for:

$$2AgCl(s) \rightleftarrows 2Ag(s) + Cl_2(g)$$

What is K_e for the process?

Answer 54

3.4×10^{-40}

OEIOU: $K_e = \dfrac{\text{antilog}\left[\dfrac{-2\Delta G°}{2.303\,RT}\right]}{(RT)^{\Delta n_g}}$

PART 16
PRECIPITATION EQUILIBRIA

EQUILIBRIUM PROBLEMS

Statement 1

When a salt, M_aX_b, is "insoluble" in water (actually, **sparingly soluble**; no salt is completely insoluble in water), mixing one solution that contains a soluble salt of the cation with another that contains a soluble salt of the anion results in **precipitation** of M_aX_b. That is, solid M_aX_b forms, leaving only a few solvated M^{+b} and X^{-a} ions behind in solution.

For example, for a salt with the formula MX_2, we may write an equation showing **equilibrium** between the precipitate* and its solvated ions:

$$MX_2 \downarrow \rightleftharpoons M^{2+} + 2X^-$$

The same equilibrium would result from placing the pure insoluble salt in water.

*Precipitates are solids and are commonly designated by placing the symbol (s) after the formula for the compound. We shall use a vertical arrow ↓ here to emphasize precipitation as a *process*.

Question 1

AgBr is insoluble in water. Write an equation for the equilibrium established when solid AgBr is placed in pure water.

Answer 1

$$AgBr \downarrow \rightleftharpoons Ag^+ + Br^-$$

Question 2

The hydroxide of Na^+ is soluble, but the hydroxide of Al^{3+} is not. The nitrates of both of these cations are soluble. Write an equilibrium equation for the precipitate formed when NaOH and $Al(NO_3)_3$ solutions are mixed.

Answer 2

$$Al^{3+} + 3OH^- \rightleftharpoons Al(OH)_3 \downarrow$$

Statement 2

By becoming familiar with those salts which may be classified as "soluble" and "insoluble," we may **predict** the **net reaction**, if any, that results from mixing solutions of two salts.

A solution of a **soluble salt** consists of solvated ions from the salt. When two such solutions are mixed, each containing one type of cation and one type of anion, two products are theoretically possible:

Solution 1: M^+ and X^-
Solution 2: M'^+ and X'^- } MX' or $M'X$ are possible products.

If either MX' or $M'X$ is **insoluble**, there will be a net reaction, shown by a **net ionic equation**:

$$M^+ + X'^- \rightarrow MX' \downarrow$$

Other ions left behind in solution are called "bystander ions" and do not appear in the equation. Information about solubility may be obtained by applying **general rules of solubility**. (See Table 16.1.)

TABLE 16.1 GENERAL RULES OF SOLUBILITY

Compounds Containing the Following Ions Are Generally Soluble in Water	Exceptions
NO_3^- $O-\overset{\underset{\parallel}{O}}{C}-CH_3$ (OAc$^-$)	AgOAc
Li$^+$ Na$^+$ K$^+$ Rb$^+$ Cs$^+$ NH$_4^+$	No common exceptions
Cl$^-$ Br$^-$ I$^-$	BiOCl SbOCl AgX Hg$_2$X$_2$ PbX$_2$ (X=Cl, Br, I)
SO_4^{2-}	PbSO$_4$ SrSO$_4$ BaSO$_4$ (CaSO$_4$ sl. sol.)
Compounds Containing the Following Ions Are Generally Insoluble	**Exceptions**
CO_3^{2-} SO_3^{2-}	Carbonates and sulfites of alkali metals and NH$_4^+$
S^{2-}	Sulfides of alkali metals and NH$_4^+$ (sulfides of alkaline earth metals, Al, and Cr *react* with water)
OH$^-$	Hydroxides of alkali metals and NH$_4^+$ (hydroxides of Ba^{2+}, Sr^{2+}, and Ca^{2+} are slightly soluble)

Question 3

Write the net ionic equation for any reaction taking place when equal volumes of 0.1 M KCl and AgNO$_3$ are mixed.

Answer 3

$$Ag^+ + Cl^- \rightarrow AgCl \downarrow$$

Question 4

Write the net ionic equation for any reaction taking place when equal volumes of 0.1 M Na$_2$S and CuSO$_4$ are mixed.

Answer 4

$$Cu^{2+} + S^{2-} \rightarrow CuS \downarrow$$

Question 5

Write the net ionic equation for any reaction taking place when equal volumes of 0.1 M Hg$_2$(NO$_3$)$_2$ and CsCl are mixed.

Answer 5

$$Hg_2^{2+} + 2Cl^- \rightarrow Hg_2Cl_2 \downarrow$$

Question 6

Write the net ionic equation for any reaction taking place when equal volumes of 0.1 M NH$_4$Br and CaCl$_2$ are mixed.

Answer 6

No reaction.

Both possible products, NH$_4$Cl and CaBr$_2$, are *soluble*.

Question 7

Write the net ionic equation for any reaction taking place when equal volumes of 0.1 M BaBr$_2$ and MgSO$_4$ solutions are mixed.

Answer 7

$$Ba^{2+} + SO_4^{2-} \rightarrow BaSO_4 \downarrow$$

Statement 3

The extent to which a sparingly soluble salt dissolves is indicated by the magnitude of the equilibrium constant, called a **solubility product constant**, associated with the equilibrium. For example, for the process:

$$MX_2 \downarrow \rightleftharpoons M^{2+} + 2X^- \qquad K_{sp} = [M^{2+}][X^-]^2$$

where the terms in brackets are equilibrium concentrations. (This relation is stirctly true only for precipitations in very dilute solutions, where interionic forces are very small, but it is sufficiently accurate for good approximations.)

The **smaller** the value of K_{sp}, the **more insoluble** the salt.

K_{sp} values may be obtained from tables for most common insoluble salts. (See Table 16.2.)

TABLE 16.2 K_{sp} VALUES

AgI	1.0×10^{-16}	NiS	1.0×10^{-22}
BaSO$_4$	1.5×10^{-9}	PbCl$_2$	1.7×10^{-5}
CaSO$_4$	3.0×10^{-5}	PbBr$_2$	5.0×10^{-6}
CuS	1×10^{-25}	PbI$_2$	1.0×10^{-8}
Hg$_2$Cl$_2$	1.3×10^{-18}	PbSO$_4$	1.6×10^{-8}
HgS	1×10^{-52}	ZnS	1.0×10^{-23}
		Ag$_2$CrO$_4$	1.9×10^{-12}

Question 8

Which is less soluble, CuS or HgS?

Answer 8

HgS

From the table we see:

$$K_{sp, \text{HgS}} < K_{sp, \text{CuS}}$$

Careful! Only for salts having the same type formula may we compare solubilities by direct comparison of K_{sp} values.

Question 9

How many moles per liter of Ag^+ and I^- go into solution when AgI is placed in pure water?

Answer 9

1.0×10^{-8} mole/L

$K_{sp} = [Ag^+][I^-]$

$= [Ag^+]^2$

OEIOU: $[Ag^+] = [I^-] = \sqrt{K_{sp}}$

$= \sqrt{1.0 \times 10^{-16}}$

$= 1.0 \times 10^{-8}$ M

Question 10

What is the solubility of AgI in pure water? (We are defining solubility here as moles/liter in solution.)

Answer 10

1.0×10^{-8} mole/L

$K_{sp} = [Ag^+][I^-]$

$= (s)(s) = s^2$

OEIOU: $s = \sqrt{K_{sp}}$

$= \sqrt{1.0 \times 10^{-16}}$

$= 1.0 \times 10^{-8}$ M

Question 11

What is the solubility of PbI_2 in pure water?

Answer 11

1.4×10^{-3} M

$$K_{sp} = [Pb^{2+}][I^-]^2$$

$$= (s)(2s)^2 = 4s^3$$

since for every mole/L of PbI_2 dissolving, 2 mole/L of I^- is produced.

OEIOU: $s = \sqrt[3]{K_{sp}/4}$

Question 12

What is the concentration of I^- in the solution in contact with solid PbI_2 in Question 11?

Answer 12

2.8×10^{-3} M

OEIOU: $[I^-] = 2s_{PbI_2}$

$$= 2(1.4 \times 10^{-3} \text{ M})$$

$$= 2.8 \times 10^{-3} \text{ M}$$

Statement 4

The **addition** from an external source of an ion common to an insoluble salt **decreases** its solubility by shifting the equilibrium in the direction of the precipitate. The **removal** of a common ion **increases** the solubility of the salt.

The exact result of the **common ion effect** can be determined by using the K_{sp} relationship to calculate the solubility of the salt, assuming that the contribution of the dissolving process to the total concentration of the common ion is negligible.

For example, when soluble NaX is added to an insoluble salt, MX, in equilibrium with its solvated ions:

$$MX \downarrow \rightleftharpoons M^+ + X^-$$

$$NaX \rightarrow Na^+ + X^-$$

$$[X^-]_{total} = [X^-]_{MX} + [X^-]_{NaX} \approx [X^-]_{NaX} = C_{NaX}$$

where C_{NaX} is the formal concentration (mole/L of added NaX).

Question 13

The solubility of lead bromide in pure water is 8.4 g/L at 20°C. Will its solubility in 1.0 M NaBr be greater or less than 8.4 g/L?

> **Answer 13**
>
> Less.
>
>> The common ion, Br^-, from NaBr will cause a shift to the left.

Question 14

The solubility of AgI in pure water is 1.0×10^{-8} M. What is its solubility in 0.50 M NaI solution?

> **Answer 14**
>
> 2.0×10^{-17} M
>
> $K_{sp} = [Ag^+][I^-]$
>
> $\approx (s_{AgI})(C_{NaI})$
>
> OEIOU: $s_{AgI} \approx K_{sp}/C_{NaI}$
>
> $= \dfrac{1.0 \times 10^{-16} \, M^2}{0.50 \, M}$
>
> $= 2.0 \times 10^{-17}$ M

Question 15

The solubility of ZnS in pure water is 3.3×10^{-12} M. If the sulfide ion concentration is reduced to 1.0×10^{-14} M by adding acid, how much ZnS will dissolve per liter?

> **Answer 15**
>
> 1.0×10^{-9} mole/L
>
> OEIOU: $s_{ZnS} = K_{sp}/[S^{2-}]$

PART 16—PRECIPITATION EQUILIBRIA

Question 16

What will be the equilibrium concentration of Br^- in solution when solid $PbBr_2$ is placed in a 0.10 M lead acetate solution?

Answer 16

7.1×10^{-3} mole/L

$$K_{sp} = [Pb^{2+}][Br^-]^2$$

$$\approx \left[C_{Pb(OAc)_2}\right][Br^-]^2$$

OEIOU: $[Br^-] \approx \sqrt{K_{sp}/C_{Pb(OAc)_2}}$

Statement 5

In order for a precipitate to form, the concentration of cation and anion must be such that Q_c, the **concentration product** when the ions are mixed, exceeds K_{sp}. For example, for:

$$MX_2 \downarrow \rightleftharpoons M^{2+} + 2X^- \qquad Q_c = [M^{2+}]_i \, [X^-]_i^2$$

where the i subscripts represent initial concentrations. Once precipitation takes place, equilibrium is established, and

$$K_{sp} = [M^{2+}]_e \, [X^-]_e^2$$

If $Q_c > K_{sp}$, a precipitate will form; otherwise no precipitation takes place.

The exact equilibrium concentration of one ion can be calculated if the K_{sp} and the concentration of the other ion are known.

Question 17

Equal volumes of solutions of $BaCl_2$ and Na_2SO_4 are mixed to give a solution that is 0.10 M in each of the two salts. Will a precipitate form?

Answer 17

Yes. $BaSO_4$ will form.

NaCl is soluble, but:

OEIOU: $Q_c = [Ba^{2+}]_i \, [SO_4^{2-}]_i$

$Q_c = 1.0 \times 10^{-2} > K_{sp} = 1.5 \times 10^{-9}$

Question 18

10.0 ml each of 0.10 M lead acetate, $Pb(OAc)_2$, and 0.10 M NaCl are mixed. Will a precipitate form?

Answer 18

Yes. $PbCl_2$ will form.

Taking into account the dilution resulting from mixing the two solutions:

$$Q_c = [Pb^{2+}]_i \, [Cl^-]_i^2$$

OEIOU: $Q_c = \left[\dfrac{M_{Pb} \times L_{Pb}}{L_{Pb} + L_{Cl}} \right] \times \left[\dfrac{M_{Cl} \times L_{Cl}}{L_{Pb} + L_{Cl}} \right]^2$

$= \left[\dfrac{(0.10 \text{ M}) (0.0100 \text{ L})}{(0.0100 + 0.0100) \text{ L}} \right] \times \left[\dfrac{(0.10 \text{ M}) (0.0100 \text{ L})}{(0.0100 + 0.0100) \text{ L}} \right]^2$

$= 1.2 \times 10^{-4}$

Thus: $Q_c = 1.2 \times 10^{-4} > K_{sp} = 1.7 \times 10^{-5}$

Question 19

500 ml each of 0.010 M $CaCl_2$ and 0.0020 M Na_2SO_4 are mixed. Will a precipitate form?

Answer 19

No.

NaCl is soluble. $CaSO_4$ has a value of 3.0×10^{-5} for its K_{sp}. Calculating Q_c:

OEIOU: $Q_c = \left[\dfrac{M_{Ca} \times L_{Ca}}{L_{Ca} + L_{SO_4}} \right] \left[\dfrac{M_{SO_4} \times L_{SO_4}}{L_{Ca} + L_{SO_4}} \right]$

$= \left[\dfrac{5.0 \times 10^{-3} \text{ mole}}{1.000 \text{ L}} \right] \times \left[\dfrac{1.0 \times 10^{-3}}{1.000 \text{ L}} \right]$

$= 5.0 \times 10^{-6}$

Thus: $Q_c = 5.0 \times 10^{-6} < K_{sp} = 3.0 \times 10^{-5}$

Question 20

A 500-ml solution is 0.10 M in $NiCl_2$. If 500 ml of 0.20 M Na_2S is added, how much Ni^{2+} is left in solution?

Answer 20

2.0×10^{-21} mole/L

$$K_{sp} = [Ni^{2+}]_e \, [S^{2-}]_e$$

$$\approx [Ni^{2+}]_e \, (C_s)$$

$$= [Ni^{2+}]_e \left(\frac{\text{moles}_{S,\,\text{initial}} - \text{moles}_{S,\,\text{react}}}{L_{\text{total}}} \right)$$

$$= [Ni^{2+}]_e \left[\frac{M_S \times L_S - M_{Ni} \times L_{Ni}}{L_{Ni} + L_S} \right]$$

OEIOU: $[Ni^{2+}]_e \approx \left[\dfrac{K_{sp} \, (L_{Ni} + L_S)}{M_S \times L_S - M_{Ni} \times L_{Ni}} \right]$

Question 21

A solution is 0.10 M in I^- and 0.10 M in SO_4^{2-}. If solid $Pb(OAc)_2$ is gradually stirred into the solution:

a. what concentration of Pb^{2+} is required for PbI_2 to precipitate?

b. what concentration of Pb^{2+} is required for $PbSO_4$ to precipitate?

c. will PbI_2 or $PbSO_4$ precipitate first?

Answer 21

a. For PbI_2: 1.0×10^{-6} M

b. For $PbSO_4$: 1.6×10^{-7} M

c. $PbSO_4$ will precipitate first.

For PbI_2 to precipitate:

$$[Pb^{2+}] = K_{sp,\,PbI_2} / [I^-]^2 \approx K_{sp}/C_{I^-}^2$$

$$= (1.0 \times 10^{-8}) / (0.10)^2$$

$$= 1.0 \times 10^{-6} \text{ M}$$

For $PbSO_4$ to precipitate:

$$[Pb^{2+}] = K_{sp,\,PbSO_4} / [SO_4^{2-}]$$

$$(1.6 \times 10^{-8})/(0.10)$$

$$= 1.6 \times 10^{-7} \text{ M}$$

A smaller concentration of Pb^{2+} is needed to precipitate $PbSO_4$, so it precipitates first.

STOICHIOMETRIC PROBLEMS

Statement 6

The formation of a precipitate can prove useful in **gravimetric analysis**. As long as the ions in equilibrium with the precipitate are present in **negligible** concentration (that is, as long as the precipitated salt is not significantly soluble) and the precipitate is otherwise suitable for handling, good results can be obtained by simply dissolving the sample of interest, adding a precipitating reagent, and **weighing** the precipitate formed.

Knowing the equation for the precipitation reaction, one then uses the stoichiometric relationship between the precipitate and constituent of interest to establish how much of the constituent must have been present in the original sample. This is usually expressed in percentage terms:

$$\% X = \frac{g\ X}{g\ \text{sample}} \times 100$$

Thus, for the general process:

$$aX + \text{reagent} \rightarrow b\ \text{ppt} \downarrow$$

the OEIOU will have the form:

$$\% X = \frac{a}{b} \times \frac{g\ \text{ppt}}{GMW_{ppt}} \times \frac{GMW_x}{g\ \text{sample}} \times 100$$

Note: There is a strong temptation to memorize equations like this. However, it is much safer to derive the equation for a given process when it is needed.

Question 22

A soluble solid is known to contain Cl^- and no other anion which would form a precipitate with Ag^+. When 0.400 g of the solid is dissolved, excess $AgNO_3$ added, and the precipitate filtered, dried, and weighed, it is found that 0.484 g of AgCl is formed. What per cent of the solid was Cl^-?

Answer 22

30.0 % Cl

Reaction: $Ag^+ + Cl^- \rightleftharpoons AgCl \downarrow$

SS: # moles Cl = # moles AgCl

PART 16—PRECIPITATION EQUILIBRIA

$$\frac{g\ Cl}{GAW_{Cl}} = \frac{g\ AgCl}{GMW_{AgCl}}$$

$$g\ Cl = \frac{g\ AgCl}{GMW_{AgCl}} \times GAW_{Cl}$$

By definition: $\%\ Cl = \frac{g\ Cl}{g\ solid} \times 100$

OEIOU: $\%\ Cl = \frac{g\ AgCl}{GMW_{AgCl}} \times \frac{GAW_{Cl}}{g\ solid} \times 100$

$$= \frac{0.484\ g}{143.3\ g} \times \frac{35.453\ g}{0.400\ g} \times 100$$

$$= 30.0\%$$

Question 23

The per cent of Ag in a sample is determined by adding excess K_2CrO_4 in order to precipitate Ag_2CrO_4. If the original sample weighs 0.800 g and the dried precipitate weighs 0.484 g, what is the per cent of Ag in the sample?

Answer 23

39.4 % Ag

Rx: $2Ag^+ + CrO_4^{2-} \rightleftharpoons Ag_2CrO_4\downarrow$

SS: # moles Ag = 2 × # moles Ag_2CrO_4

OEIOU: $\%\ Ag = 2 \times \frac{g\ Ag_2CrO_4}{GMW_{Ag_2CrO_4}} \times \frac{GAW_{Ag}}{g\ sample} \times 100$

Statement 7

Volumetric analysis involving precipitate formation is also possible. This differs from gravimetric analysis in that one **titrates**—that is, one adds increments of a precipitating solution—until exactly enough reagent has been added (as shown by a suitable indicator) to precipitate all the ion of interest.

From the proper stoichiometric statement and a knowledge of the concentration and added volume of the titrant, one may then calculate the percentage of the "unknown" in the original sample.

For the general process: $aX + b$ titrant $\rightarrow$ ppt $\downarrow$

the OEIOU will have the form:

$$\% X = \frac{a}{b} \times (M \times \text{liter})_{\text{Titr}} \times \frac{\text{GMW}_X}{\text{g sample}} \times 100$$

As in Statement 6, it is advisable to avoid memorizing. Derive the proper expression from a correct stoichiometric statement when you need it.

Question 24

A soluble solid contains Cl^- and no other ion that forms a precipitate with Ag^+. If 0.400 g of the solid is dissolved, and the solution is titrated with 0.100 M $AgNO_3$, it is found that 33.80 ml is required to precipitate all the Cl^- as AgCl. What is the per cent of Cl in the solid?

Answer 24

30.0 % Cl

Rx: $Ag^+ + Cl^- \rightarrow AgCl\downarrow$

SS: # moles Cl = # moles Ag = # moles AgCl

Substituting: $\dfrac{\text{g Ag}}{\text{GAW}_{Ag}} = (M \times L)_{AgNO_3}$

OEIOU: $\% Cl = (M \times L)_{AgNO_3} \times \dfrac{\text{GAW}_{Cl}}{\text{g solid}} \times 100$

$= (0.100 \text{ M} \times 0.03380 \text{ L}) \times \dfrac{35.453 \text{ g}}{0.400 \text{ g}} \times 100$

$= 30.0 \%$

Question 25

If 14.60 ml of 0.100 M K_2CrO_4 is required to precipitate all the Ag^+ in a 0.800-g sample of a soluble salt, what is the per cent of Ag in the salt?

Answer 25

39.4 % Ag

Rx: $2Ag^+ + CrO_4^{2-} \rightarrow Ag_2CrO_4 \downarrow$

SS: # moles Ag = 2 × # moles CrO_4^{2-}

$$= 2 \times \text{\# moles } K_2CrO_4$$

$$\text{Thus:} \quad \frac{\text{g Ag}}{\text{GAW}_{Ag}} = 2 \times (M \times L)_{K_2CrO_4}$$

$$\text{OEIOU:} \quad \% \text{ Ag} = 2\,(M \times L)_{K_2CrO_4} \times \frac{\text{GAW}_{Ag}}{\text{g sample}} \times 100$$

TESTING YOUR MASTERY—Part 16
Precipitation Equilibria

Question 26

Write the net ionic equation for any reaction taking place when equal volumes of 0.1 M solutions of these salts are mixed:

a. CsCl and NH_4NO_3

b. $Mg(NO_3)_2$ and NaOH

c. K_2S and $NiBr_2$

d. $Pb(OAc)_2$ and Li_2SO_4

Answer 26

a. no reaction

b. $Mg^{2+} + 2OH^- \rightarrow Mg(OH)_2 \downarrow$

c. $Ni^{2+} + S^{2-} \rightarrow NiS \downarrow$

d. $Pb^{2+} + SO_4^{2-} \rightarrow PbSO_4 \downarrow$

Question 27

Arrange these lead salts in order of decreasing solubility:

$$PbBr_2 \quad PbCl_2 \quad PbI_2$$

Answer 27

$$PbCl_2 > PbBr_2 > PbI_2$$

Because the formulas are similar, the salts are arranged in order of decreasing K_{sp}.

Question 28

Which of these is more soluble: PbI_2 or $PbSO_4$?

Answer 28

PbI_2 is more soluble.

Although $K_{sp,PbSO_4} < K_{sp,PbI_2}$, solubility and K_{sp} are related differently for these salts having unlike formulas.

for PbI_2: $s_{PbI_2} = 3\sqrt{K_{sp}/4}$

for $PbSO_4$: $s_{PbSO_4} = \sqrt{K_{sp}}$

As it turns out, $s_{PbSO_4} < s_{PbI_2}$

Question 29

What is the solubility of $PbCl_2$ in water expressed in:

a. mole/liter?

b. g/liter?

Answer 29

a. 1.6×10^{-2} mole/liter

OEIOU: $s = \sqrt[3]{\dfrac{K_{sp}}{4}}$

b. 4.4 g/liter

OEIOU: $\dfrac{g}{L} = \sqrt[3]{\dfrac{K_{sp}}{4}} \times (GMW)_{PbCl_2}$

Question 30

A reasonably large amount of solid Ag_2CrO_4 is stirred with pure water.

a. How many moles of Ag_2CrO_4 will dissolve per liter?

b. What will be the molar concentration of Ag^+ in the resulting solution?

Answer 30

a. $s = 7.8 \times 10^{-5}$ mole/liter

OEIOU: $s = \sqrt[3]{\dfrac{K_{sp}}{4}}$

b. $[Ag^+] = 1.6 \times 10^{-4}$ mole/liter

OEIOU: $[Ag^+] = 2s$

Question 31

What would be the molar concentration of Hg^+ in a solution formed by stirring an excess of solid HgS in water?

Answer 31

1.0×10^{-26} mole/liter

OEIOU: $[Hg^+] = \sqrt{K_{sp}}$

Question 32

What is the expected solubility of ZnS in:

a. pure water?

b. a 0.010 M Na_2S solution?

Answer 32

a. 3.2×10^{-12} M

OEIOU: $s = \sqrt{K_{sp,\,ZnS}}$

b. 1.0×10^{-21} M

OEIOU: $s \approx \dfrac{K_{sp,\,ZnS}}{C_{Na_2S}}$

Question 33

What is the solubility in a 0.080 M $AgNO_3$ solution of:

a. AgI?

b. Ag_2CrO_4?

Answer 33

a. 1.6×10^{-14} M

OEIOU: $s \approx \dfrac{K_{sp,\,AgI}}{C_{AgNO_3}}$

b. 3.0×10^{-10} M

OEIOU: $s \approx \dfrac{K_{sp,\,Ag_2CrO_4}}{4\,C^2_{AgNO_3}}$

Question 34

What is the solubility of $CaSO_4$ in a 0.010 M $NaSO_4$ solution?

Answer 34

2.5×10^{-3} mole/L

OEIOU: $K_{sp} = s(s + C_{Na_2SO_4})$

Solve for s.

The difference between s and $C_{Na_2SO_4}$ is not great enough here to ignore s in the $(s + C_{Na_2SO_4})$ term.

Question 35

What will be the equilibrium concentration of Ag^+ in solution when solid Ag_2CrO_4 is stirred with a 0.085 M K_2CrO_4 solution?

Answer 35

4.7×10^{-6} M

OEIOU: $[Ag^+] \approx \sqrt{\dfrac{K_{sp,\,Ag_2CrO_4}}{C_{K_2CrO_4}}}$

Question 36

The solubility of $SrCO_3$ in pure water is 2.6×10^{-5} M.

a. What is the K_{sp} of $SrCO_3$?

b. What is the solubility of $SrCO_3$ if the CO_3^{2-} concentration is reduced to 1.3×10^{-8} M by adding acid?

c. Would the addition of a strong acid be a good method of dissolving $SrCO_3$?

Answer 36

a. 6.8×10^{-10} M

OEIOU: $K_{sp} = s^2{}_{SrCO_3}$

b. 5.2×10^{-2} M

OEIOU: $s_{SrCO_3} = \dfrac{K_{sp}}{[CO_3^{2-}]}$

c. Yes.

The solubility is greatly increased by the addition of acid. Compare 2.6×10^{-5} M (in pure water) with 5.2×10^{-2} M (in the acid solution).

Question 37

Solutions of $CaCl_2$ and K_2SO_4 are mixed to give a solution that is 0.018 M in each of the salts. Will a precipitate form?

Answer 37

Yes, because $Q_c > K_{sp, CaSO_4}$

OEIOU: $Q_c = [Ca^{2+}]_i \, [SO_4^{2-}]_i$

Question 38

If 100.0 ml each of 0.10 M $AgNO_3$ and 0.010 M Na_2CrO_4 are mixed, will a precipitate form?

Answer 38

Yes.

$Q_c = 1.3 \times 10^{-5} > K_{sp} = 1.9 \times 10^{-12}$

OEIOU: $Q_c = \left[\dfrac{M_{Ag} \times L_{Ag}}{L_{Ag} + L_{CrO_4}}\right]^2 \left[\dfrac{M_{CrO_4} \times L_{CrO_4}}{L_{Ag} + L_{CrO_4}}\right]$

Question 39

A solution of 0.0010 M AgNO$_3$ has a volume of 200.0 ml. If a KI solution is to be added to it, what minimum concentration must it have to give a precipitate of AgI when the total volume is 500.0 ml?

Answer 39

4.2 × 10^{-13} M

$$\text{OEIOU: } M_{\text{NaI}} = \left[\frac{L_{\text{total}}^2}{M_{\text{Ag}} \times L_{\text{Ag}}}\right] \times \left[\frac{K_{sp}}{(L_{\text{total}} - L_{\text{Ag}})}\right]$$

Question 40

A solution is 0.010 M in I$^-$ and CrO$_4^{2-}$. If solid AgNO$_3$ is gradually stirred into the solution, will Ag$_2$CrO$_4$ or AgI precipitate first?

Answer 40

AgI will precipitate first, because less Ag$^+$ is required to allow $Q_{\text{AgI}} > K_{sp,\text{AgI}}$ than is needed to allow $Q_{\text{Ag}_2\text{CrO}_4} > K_{sp,\text{Ag}_2\text{CrO}_4}$.

For AgI, the *minimum* [Ag$^+$] is found from:

$$\text{OEIOU: } [\text{Ag}^+] \approx \frac{K_{sp,\text{AgI}}}{C_I}$$

For Ag$_2$CrO$_4$:

$$\text{OEIOU: } [\text{Ag}^+] \approx \sqrt{\frac{K_{sp,\text{Ag}_2\text{CrO}_4}}{C_{\text{CrO}_4}}}$$

Question 41

A 600-ml solution of 0.10 M Pb(NO$_3$)$_2$ and 400-ml of 0.30 M of Na$_2$SO$_4$ are mixed. How much Pb^{2+} is left in solution?

Answer 41

2.7 × 10^{-7} M

OEIOU: $[Pb^{2+}] \approx \dfrac{K_{sp,PbSO_4}(L_{Pb} + L_{SO_4})}{(M \times L)_{SO_4} - (M \times L)_{Pb}}$

Question 42

A soluble impure sulfate salt is titrated with 0.108 M $BaCl_2$ solution. If 20.13 ml of $BaCl_2$ solution was required to titrate a 0.500-g sample of the salt, what was the per cent of SO_4 present?

Answer 42

44.2% SO_4

OEIOU: $\% SO_4 = \dfrac{M_{Ba} \times L_{Ba} \times (GMW)_{SO_4}}{\text{g salt}} \times 100$

Question 43

The per cent of I in a certain soluble salt is determined by adding excess $Pb(OAc)_2$ and weighing the dried PbI_2 formed. If 0.741 g of PbI_2 is obtained from a 0.800-g sample of the salt, what is the per cent of I in the salt?

Answer 43

51.0 %

OEIOU: $\% I = 2 \times \left(\dfrac{g}{GMW}\right)_{PbI_2} \times \dfrac{(GAW)_I}{g_{salt}} \times 100$

PART 17
ACIDS AND BASES

Statement 1

Water undergoes **autoionization** to give H^+ and OH^- (or H_3O^+ and OH^-; we will use the H^+ symbol, even though this species does not exist free in water), so that we may write—for pure water or for any aqueous solution—the equilibrium:

$$H_2O \rightleftharpoons H^+ + OH^-$$

The slight extent to which the ionization takes place is indicated by the small value for the **ion product constant** for water, K_w:

$$K_w = [H^+][OH^-] = 1.0 \times 10^{-14}$$

When H^+ or OH^- ions are also supplied from some other source, such as addition of an acid or base to the water, the concentration of H^+ may exceed the concentration of OH^-, or vice versa. In every case, however, the K_w relationship is applicable.

Question 1

What is the concentration of H^+ in pure water? What is the concentration of OH^-?

Answer 1

$[H^+] = 1.0 \times 10^{-7}$ mole/L

$[OH^-] = 1.0 \times 10^{-7}$ mole/L

$$K_w = [H^+][OH^-]$$
$$= [H^+]^2$$

OEIOU: $[H^+] = [OH^-] = \sqrt{K_w}$

$$= \sqrt{1.0 \times 10^{-14}}$$
$$= 1.0 \times 10^{-7} \text{ M}$$

Question 2

What is the concentration of OH^- in an aqueous solution having an H^+ concentration of 1.0×10^{-2} M?

Answer 2

1.0×10^{-12} M

$K_w = [H^+][OH^-]$

OEIOU: $[OH^-] = K_w/[H^+]$

Question 3

What is the concentration of H^+ ion in an aqueous solution having an OH^- concentration of 1.0×10^{-2} M?

Answer 3

1.0×10^{-12} M

$K_w = [H^+][OH^-]$

OEIOU: $[H^+] = K_w/[OH^-]$

Statement 2

The hydrogen ion content of a solution may also be expressed in **pH units**, where:

$$pH = -\log[H^+]$$

Similarly:

$$pOH = -\log[OH^-]$$

From the K_w relationship:

$$pH + pOH = 14.00 \ldots$$

(You may wish to refer again to Statements 15 and 21 in Part 1 to review logarithms.)

Question 4

What is the pH of a solution in which the hydrogen ion concentration is 1.0×10^{-9} M?

Answer 4

9.0

OEIOU: $pH = -\log [H^+]$
$= -\log (1.0 \times 10^{-9})$
$= -(-9.0)$
$= 9.0$

Question 5

What are the pH and pOH values for pure water?

Answer 5

$pH = pOH = 7.0$

OEIOU: $pH = pOH = -\log \sqrt{K_w}$

Question 6

What are the pH and pOH of a solution having a hydrogen ion concentration of 2.5×10^{-6} M?

Answer 6

$pH = 5.6 \qquad pOH = 8.4$

OEIOU: $pH = -\log [H^+]$
$= -\log (2.5 \times 10^{-6})$
$= 5.6$

OEIOU: $pOH = 14.0 - pH$

Question 7

A certain solution has a pH of 1.5.

a. What is the hydrogen ion concentration of the solution?

b. What is the OH⁻ concentration?

Answer 7

a. 3.2×10^{-2} M

b. 3.1×10^{-13} M

$$pH = -\log [H^+]$$

$$\text{OEIOU:} \quad [H^+] = \text{antilog}(-pH)$$
$$= \text{antilog}(-1.5)$$
$$= 3.2 \times 10^{-2}$$

$$\text{OEIOU:} \quad [OH^-] = \frac{K_w}{\text{antilog}(-pH)}$$

Question 8

A solution having a hydrogen ion concentration of greater than 10^{-7} M is said to be *acidic*; one with a H^+ concentration of less than 10^{-7} is said to be *basic*. What are the pH limits for acidity and basicity?

Answer 8

acidic: pH < 7

basic: pH > 7

Statement 3

When a **strong acid**, HX, is dissolved in water, it undergoes essentially **complete ionization**:

$$HX \rightarrow H^+ + X^-$$

In a solution of a strong acid, then (except at extremely low concentration), the H^+ from the ionization of water will be negligible, with most of the H^+ being supplied by the strong acid:

$$[H^+]_{total} = [H^+]_{HX} + [H^+]_{H_2O} \approx [H^+]_{HX}$$

Or, if C_{HX} represents the "**formal concentration**" of HX (that is, the total moles of dissolved HX per liter, regardless of the form to which it converts in solution):

$$[H^+] \approx C_{HX}$$

Question 9

What is the hydrogen ion concentration in a 6.0 M hydrochloric acid solution?

Answer 9

6.0 M

OEIOU: $[H^+] \approx C_{HCl}$

$= 6.0$ M

Question 10

What is the hydrogen ion concentration in a solution formed by dissolving 63 g of HNO_3 in 2.0 liters of water?

Answer 10

0.50 M

$[H^+] \approx C_{HNO_3}$

$= \#\text{ moles } HNO_3/L_{solution}$

OEIOU: $[H^+] \approx (g\ HNO_3/GMW_{HNO_3})/L_{solution}$

Question 11

What is the OH^- concentration in a 0.50 M HNO_3 solution?

Answer 11

2.0×10^{-14} M

$K_w = [H^+][OH^-]$

$\approx C_{HNO_3}[OH^-]$

OEIOU: $[OH^-] \approx K_w/C_{HNO_3}$

Question 12

Calculate the concentrations of all molecular and ionic species (except H_2O) in a 0.20 M solution of HBr.

Answer 12

$[H^+] \approx 0.20$ M

$[OH^-] \approx 5.0 \times 10^{-14}$ M

$[Br^-] = 0.20$ M

First, decide what species are present:

$$HBr \rightarrow H^+ + Br^-$$

Only H^+ and Br^- present after HBr dissociates.

$$H_2O \rightleftharpoons H^+ + OH^-$$

H_2O, H^+, and OH^- all present.

Then: OEIOU: $[H^+] \approx C_{HBr}$

OEIOU: $[OH^-] \approx K_w/C_{HBr}$

OEIOU: $[Br^-] = C_{HBr}$

Statement 4

When a strong base, MOH*, is dissolved in water, it undergoes essentially **complete ionization**:

$$MOH \rightarrow M^+ + OH^-$$

Thus, in a solution of a strong base (except at extremely low concentrations), if C_{MOH} is the formal concentration of MOH:

$$[OH^-] \approx C_{MOH}$$

*In this text we will use MOH to denote a strong base; note that there are bases with more than one hydroxide per formula, e.g., $M(OH)_2$.

Question 13

What are the OH^- and H^+ concentrations in a 6.0 M NaOH solution?

Answer 13

$[OH^-] \approx 6.0$ M

$[H^+] \approx 1.7 \times 10^{-15}$ M

OEIOU: $[OH^-] \approx C_{NaOH}$

OEIOU: $[H^+] \approx K_w/C_{NaOH}$

Question 14

What is the hydroxide ion concentration in a 0.30 M Ba(OH)$_2$ solution?

Answer 14

0.60 M

Note that for this base there are 2 OH$^-$ produced per formula of base:

$$Ba(OH)_2 \rightarrow Ba^{2+} + 2OH^-$$

Thus: $[OH^-] \approx 2C_{MOH}$

Question 15

Calculate the concentrations of all molecular and ionic species (except H$_2$O) in a 0.20 M solution of potassium hydroxide.

Answer 15

$[OH^-] \approx 0.20$ M

$[H^+] \approx 5.0 \times 10^{-14}$ M

$[K^+] \doteq 0.20$ M

OEIOU: $[OH^-] \approx C_{MOH}$

OEIOU: $[H^+] \approx K_w/C_{MOH}$

OEIOU: $[K^+] = C_{MOH}$

Compare with Answer 12 above.

Statement 5

When a **weak monoacid**, HA, (only one reactive H per molecule) is dissolved in water, it undergoes only **partial ionization**, with most of the acid remaining in the molecular form:

$$HA \rightleftharpoons H^+ + A^-$$

The **acidity constant**, K_a, indicates the extent of this ionization:

$$K_a = \frac{[H^+][A^-]}{[HA]}$$

Except for extremely weak acids or extremely low concentrations, most of the H^+ in solution still comes from the acid rather than from water, and $[H^+] \approx [A^-]$. If C_A is the formal concentration of the acid (that is, the total moles of dissolved acid per liter, both ionized and un-ionized), the concentration of un-ionized (molecular) HA will be given by: $[HA] \approx C_A - [H^+]$. Thus:

$$K_a \approx \frac{[H^+]^2}{C_A - [H^+]}$$

When K_a and C_A are known, we can calculate $[H^+]$ from this equation.

When $H^+ < 0.05\ C_A$, as is often the case, a further approximation is usually justified:

$$K_a \approx \frac{[H^+]^2}{C_A} \quad \text{and} \quad [H^+] \approx \sqrt{K_a C_A}$$

(When it is unclear whether $[H^+] < 0.05\ C_A$, assume that it is, solve for $[H^+]$, and see whether the assumption was valid. If it is not, one must solve the more complicated quadratic equation.) K_a values are listed in Table 17.1.

TABLE 17.1 K_a VALUES AT 25°C

$HClO_2$	1.1×10^{-2}	HOCl	3.2×10^{-8}
H_3PO_4	7.5×10^{-3}	HCN	7.2×10^{-10}
HF	7.0×10^{-4}	NH_4^+	5.6×10^{-10}
HOAc	1.8×10^{-5}	HCO_3^-	4.8×10^{-11}
H_2S	1.1×10^{-7}	HPO_4^{2-}	1.0×10^{-12}
$H_2PO_4^-$	6.2×10^{-8}	HS^-	1.0×10^{-14}

Question 16

What is the hydrogen ion concentration in a 0.10 M acetic acid solution?

Answer 16

1.3×10^{-3} M

OEIOU: $[H^+] \approx \sqrt{K_a C_A}$

$= \sqrt{(1.8 \times 10^{-5})(0.10)}$

$= 1.3 \times 10^{-3}$ M

Question 17

What are the hydrogen ion and hydroxide ion concentrations in a solution of 0.40 M hypochlorous acid?

Answer 17

$[H^+] \approx 1.1 \times 10^{-4}$ M

$[OH^-] \approx 9.1 \times 10^{-11}$ M

OEIOU: $[H^+] \approx \sqrt{K_a C_A}$

OEIOU: $[OH^-] \approx K_w / \sqrt{K_a C_A}$

Question 18

What are the H^+ ion and OH^- ion concentraions in a solution of 0.40 M chlorous acid?

Answer 18

$[H^+] \approx 0.061$ M

$[OH^-] \approx 1.6 \times 10^{-13}$

Testing: $[H^+] \approx \sqrt{K_a C_A}$

$= \sqrt{4.4 \times 10^{-3}}$

$= 0.066$ M

Since $[H^+] \approx 0.066 > 0.02 = 0.05\, C_A$, we must solve the more complicated quadratic equation:

$$K_a \approx \frac{[H^+]^2}{C_A - [H^+]}$$

or: $[H^+]^2 + K_a [H^+] - K_a C_A \approx 0$

PART 17—ACIDS AND BASES

$$\text{OEIOU:} \quad [H^+] \approx \frac{-K_a \pm \sqrt{K_a^2 - 4(K_a C_A)}}{2}$$

$$= \frac{-(1.1 \times 10^{-2}) \pm (1.3 \times 10^{-1})}{2}$$

$$= 0.060 \text{ M} \quad \text{or} \quad -0.071 \text{ M}$$

Negative concentrations do not exist, so 0.060 M is the only possible answer.

OEIOU: $[OH^-] \approx K_w/[H^+]$ from above quadratic solution

---

Question 19

Calculate the concentrations of all molecular and ionic species (except H_2O) in a 0.30 M HF solution.

Answer 19

$[H^+] \approx 1.4 \times 10^{-2}$ M

$[OH^-] \approx 7.1 \times 10^{-13}$ M

$[HF] \approx 0.30$ M

$[F^-] \approx 1.4 \times 10^{-2}$ M

OEIOU: $[H^+] \approx [F^-] \approx \sqrt{K_a C_A}$

OEIOU: $[HF] \approx C_A$

OEIOU: $[OH^-] \approx K_w/\sqrt{K_a C_A}$

Question 20

The hydrogen ion concentration in a 0.10 M solution of a certain weak acid is 1.0×10^{-4} M. What is K_a for the acid?

Answer 20

1.0×10^{-7} M

OEIOU: $K_a \approx [H^+]^2/C_A$

Question 21

What is the pH of a vinegar salad dressing in which acetic acid is 0.050 M?

Answer 21

3.02

$$pH = -\log [H^+]$$

OEIOU: $pH \approx -\log \sqrt{K_a C_A}$

Statement 6

A **weak monobase** (a species that can accept only one proton from an acid) may be either an anion, A^- (from a salt, MA), or a molecule, B. The A^- type, the conjugate base of a weak molecular acid, is more common in aqueous solution. In either case, the base undergoes **partial ionization** to give an OH^- ion and its conjugate acid:

$$\underset{\text{base}}{A^-} + H_2O \rightleftharpoons \underset{\text{acid}}{HA} + OH^- \quad (\text{or } \underset{\text{base}}{B} + H_2O \rightleftharpoons \underset{\text{acid}}{BH^+} + OH^-)$$

Most of the base remains in its original form (A^- or B). The **basicity constant**, K_b, indicates the extent of ionization:

$$K_b = \frac{[HA][OH^-]}{[A^-]} \quad \left(\text{or } K_b = \frac{[BH^+][OH^-]}{[B]} \right)$$

Except for extremely weak bases or extremely low concentrations, most of the OH^- comes from the base rather than from the ionization of water, and $[OH^-] \approx [HA]$. Thus:

$$K_b \approx \frac{[OH^-]^2}{C_B - [OH^-]}$$

where C_B is the formal concentration of the weak base.

When C_B and K_b are known, we can derive $[OH^-]$ from the equation. And when $[OH^-] < 0.05\, C_B$:

$$K_b \approx \frac{[OH^-]^2}{C_B} \quad \text{and} \quad [OH^-] \approx \sqrt{K_b C_B}$$

K_b values are listed in Table 17.2. (When it is unclear whether $[OH^-]$ is less than $0.05\, C_B$, assume that it is, solve for $[OH^-]$, and see whether the assumption was valid. If it is not, the more complicated quadratic equation must be solved.)

K_b is related to K_a for its conjugate acid by the following equation: $K_a K_b = K_w$

PART 17—ACIDS AND BASES

TABLE 17.2 K_b VALUES AT 25°C

CO_3^{2-}	2.1×10^{-4}	HCO_3^-	2.4×10^{-8}
CN^-	2.5×10^{-5}	OAc^-	5.6×10^{-10}
NH_3	1.8×10^{-5}	F^-	1.4×10^{-11}
ClO^-	3.1×10^{-7}		

Question 22

What is the hydroxide ion concentration in a 0.10 M solution of NaCN?

Answer 22

1.6×10^{-3} M

NaCN is a salt that dissolves to give the base CN^-:

$$NaCN \rightarrow Na^+ + CN^-$$
$$\text{base}$$

OEIOU: $[OH^-] \approx \sqrt{K_b C_B}$

$\qquad = \sqrt{(2.5 \times 10^{-5})(0.10)}$

$\qquad = 1.6 \times 10^{-3}$ M

Question 23

What are the concentrations of H^+ and OH^- in a solution of 0.20 M sodium hypochlorite?

Answer 23

$[OH^-] \approx 2.5 \times 10^{-4}$ M

$[H^+] \approx 4.0 \times 10^{-11}$ M

$$NaClO \rightarrow Na^+ + ClO^-$$
$$\text{base}$$

OEIOU: $[OH^-] \approx \sqrt{K_b C_B}$

OEIOU: $[H^+] \approx K_w / \sqrt{K_b C_B}$

Question 24

Calculate the concentrations of all molecular and ionic species (except H_2O) in a solution of 0.30 M NH_3.

Answer 24

$[OH^-] \approx 2.3 \times 10^{-3}$ M

$[H^+] \approx 4.3 \times 10^{-12}$ M

$[NH_3] \approx 0.30$ M

$[NH_4^+] \approx 2.3 \times 10^{-3}$ M

OEIOU: $[OH^-] \approx [NH_4^+] \approx \sqrt{K_b C_B}$

OEIOU: $[H^+] \approx K_w/\sqrt{K_b C_B}$

OEIOU: $[NH_3] \approx C_B$

(Compare with Question 19.)

Question 25

What is the pOH of a 0.10 M sodium acetate solution?

Answer 25

5.1

$NaOAc \rightarrow Na^+ + OAc^-$
                            base

OEIOU: $pOH \approx -\log\sqrt{K_b C_B}$

Question 26

The hydroxide ion concentration of a certain 0.50 M weak base solution is 2.5×10^{-4} M. What is its K_b?

Answer 26

1.3×10^{-7}

OEIOU: $K_b \approx \dfrac{[OH^-]^2}{C_B}$

Statement 7

A **polyacid** has more than one acid hydrogen per molecule and undergoes stepwise ionization, each step of which has a characteristic K_a value. For a diacid, for instance:

$$H_2A \rightleftharpoons H^+ + HA^- \qquad K_{a_1} = \frac{[H^+][HA^-]}{[H_2A]}$$

$$HA^- \rightleftharpoons H^+ + A^{2-} \qquad K_{a_2} = \frac{[H^+][A^{2-}]}{[HA^-]}$$

As with monoacids, most of the H^+ comes from the ionization of the acid; in fact, since K_{a_2} is usually much smaller than K_{a_1}, most H^+ comes from the first ionization step. Thus for polyacids, as for monoacids, $[H_2A] \approx C_A - [H^+]$, and:

$$K_{a_1} \approx \frac{[H^+]^2}{C_A - [H^+]}$$

And when $[H^+] < 0.05 \, C_A$:

$$[H^+] \approx \sqrt{K_a C_A}$$

In addition, if only a small quantity of HA^- ionizes (that is, if $K_{a_2} \ll K_{a_1}$):

$$[HA^-] \approx [H^+]$$

$$[A^{2-}] \approx K_{a_2}$$

Question 27

Calculate the pH and pOH of a 0.080 M H_2S solution.

Answer 27

pH ≈ 4.0

pOH ≈ 10.0

OEIOU: $pH \approx -\log \sqrt{K_{a_1} C_A}$

$= -\log \sqrt{(1.1 \times 10^{-7})(8.0 \times 10^{-2})}$

$= -\log \sqrt{8.8 \times 10^{-9}}$

$= -\log (9.3 \times 10^{-5})$

$= -\log (-4.0)$

$= 4.0$

OEIOU: $pOH \approx 14.0 + \log \sqrt{K_{a_1} C_a}$

Question 28

Calculate the concentrations of all molecular and ionic species in a 0.10 M hydrosulfuric acid (or hydrogen sulfide) solution.

Answer 28

$[H^+] \approx 1.0 \times 10^{-4}$ M

$[OH^-] \approx 1.0 \times 10^{-10}$ M

$[H_2S] \approx 0.10$ M

$[HS^-] \approx 1.0 \times 10^{-4}$ M

$[S^{2-}] \approx 1.0 \times 10^{-14}$ M

Ionization steps:

$$H_2S \rightleftharpoons H^+ + HS^-$$

$$HS^- \rightleftharpoons H^+ + S^{2-}$$

Then: OEIOU: $[H^+] \approx \sqrt{K_{a_1} C_A}$

OEIOU: $[OH^-] \approx K_w/\sqrt{K_{a_1} C_A}$

OEIOU: $[H_2S] \approx C_A$

OEIOU: $[HS^-] \approx \sqrt{K_{a_1} C_A}$

OEIOU: $[S^{2-}] \approx K_{a_2}$

Question 29

Calculate the concentrations of all molecular and ionic species in a 0.080 M phosphoric acid solution.

Answer 29

$[H^+] \approx 2.1 \times 10^{-2}$ M

$[OH^-] \approx 4.8 \times 10^{-13}$ M

$[H_3PO_4] \approx 5.9 \times 10^{-2}$ M

$[H_2PO_4^-] \approx 2.1 \times 10^{-2}$ M

$[HPO_4^{2-}] \approx 6.2 \times 10^{-8}$ M

$$[PO_4^{3-}] \approx 3.0 \times 10^{-18} \text{ M}$$

Ionization steps:
$$H_3PO_4 \rightleftharpoons H_2PO_4^- + H^+$$
$$H_2PO_4^- \rightleftharpoons HPO_4^{2-} + H^+$$
$$HPO_4^{2-} \rightleftharpoons PO_4^{3-} + H^+$$

OEIOU: $K_{a_1} \approx \dfrac{[H^+]^2}{C_A - [H^+]}$

Solve for $[H^+]$.

OEIOU: $[OH^-] \approx K_w/[H^+]$ from quadratic equation

OEIOU: $[H_3PO_4] \approx C_A - [H^+]$ from quadratic equation

OEIOU: $[H_2PO_4^-] \approx [H^+]$ from quadratic equation

OEIOU: $[HPO_4^{2-}] \approx K_{a_2}$

OEIOU: $[PO_4^{3-}] \approx \dfrac{K_{a_3} K_{a_2}}{[H^+]}$ from quadratic equation

---

Statement 8

A **polybase** typically is of the form A^{2-} or A^{3-} (from the salt M_2A or M_3A), although a few bases are neutral molecules, B-B, having two (or more) points at which H^+ may attach.

For the typical A^{2-} type, we may show a stepwise reaction with water:

$$A^{2-} + H_2O \rightleftharpoons HA^- + OH^- \qquad K_{b_1} = \dfrac{[HA^-][OH^-]}{[A^{2-}]}$$

$$HA^- + H_2O \rightleftharpoons H_2A + OH^- \qquad K_{b_2} = \dfrac{[H_2A][OH^-]}{[HA^-]}$$

As with monobases, most of the OH^- comes from the ionization of base; in fact, since $K_2 \ll K_1$, most of it comes from the first ionization step. Thus:

$$[A^{2-}] \approx C_B - [OH^-]$$

$$K_{b_1} \approx \dfrac{[OH^-]^2}{C_B - [OH^-]}$$

And when $[OH^-] < 0.05\ C_B$:

$$[OH^-] \approx \sqrt{K_b C_B}$$

Also, if not much HA^- ionizes:

$$[HA^-] \approx [OH^-]$$

and:

$$[H_2A] \approx K_{b_2}$$

Question 30

Calculate the pH and pOH of a 0.10 M Na_2CO_3 solution.

Answer 30

$pH \approx 11.7 \qquad pOH \approx 2.3$

OEIOU: $pOH \approx -\log \sqrt{K_{b_1} C_B}$

$\qquad\qquad\quad = -\log \sqrt{(2.1 \times 10^{-4})(0.10)}$

$\qquad\qquad\quad = -\log \sqrt{2.1 \times 10^{-5}}$

$\qquad\qquad\quad = -\log(-2.3)$

$\qquad\qquad\quad = 2.3$

OEIOU: $pH \approx 14.0 + \log \sqrt{K_{b_1} C_B}$

Question 31

Calculate the concentrations of all molecular and ionic species in a 0.10 M K_2CO_3 solution.

Answer 31

$[OH^-]$	$\approx 4.6 \times 10^{-3}$ M
$[H^+]$	$\approx 2.2 \times 10^{-12}$ M
$[CO_3^{2-}]$	≈ 0.10
$[HCO_3^-]$	$\approx 4.6 \times 10^{-3}$ M
$[H_2CO_3]$	$\approx 2.2 \times 10^{-8}$ M
$[K^+]$	≈ 0.20 M

Ionization steps:

$$CO_3^{2-} + H_2O \rightleftharpoons HCO_3^- + OH^-$$

$$HCO_3^- + H_2O \rightleftharpoons H_2CO_3 + OH^-$$

(H_2CO_3 actually decomposes to give CO_2 and H_2O, but we won't worry about this now.)

OEIOU: $[OH^-] \approx \sqrt{K_{b_1} C_B}$

OEIOU: $[H^+] \approx K_w/\sqrt{K_{b_1} C_B}$

OEIOU: $[CO_3^{2-}] \approx C_B$

OEIOU: $[HCO_3^-] \approx \sqrt{K_{b_1} C_B}$

OEIOU: $[H_2CO_3] \approx K_{b_2}$

OEIOU: $[K^+] = 2C_B$ because $K_2CO_3 \rightarrow 2K^+ + CO_3^{2-}$

Statement 9

Salts are usually categorized according to the types of acids and bases which, in theory, could have reacted to produce them.

The salt of a **weak acid** and **strong base**, MA, will produce a **basic** solution by hydrolysis (reaction with water):

$$MA \rightarrow M^+ + A^- \qquad A^- + H_2O \rightleftharpoons HA + OH^- \qquad pH > 7$$

The salt of a **strong acid** and **weak base**, BHX, will produce an **acidic** solution by hydrolysis:

$$BHX \rightarrow BH^+ + X^- \qquad BH^+ + H_2O \rightleftharpoons B + H_3O^+ \qquad pH < 7$$

The salt of a **strong acid** and **strong base**, MX, undergoes no hydrolysis and produces a **neutral** solution.

$$MX \rightarrow M^+ + X^- \qquad X^- + H_2O \rightarrow \text{no reaction}$$

The salt of a **weak acid** and **weak base**, BHA, undergoes both basic and acidic hydrolysis:

$$BHA \rightarrow BH^+ + A^- \qquad BH^+ + H_2O \rightleftharpoons B + H_3O^+$$

$$A^- + H_2O \rightleftharpoons HA + OH^-$$

The hydrogen ion concentration of such an "**ampholyte**" solution is given by:

$$[H^+] \approx \sqrt{K_a K_w / K_b}$$

If $K_{a,BH^+} > K_{b, A^-}$, the solution is **acidic**.

If $K_{a, BH^+} < K_{b, A^+}$, the solution is **basic**.

A salt of the type MHA is also an ampholyte:

$$HA^- + H_2O \rightleftharpoons A^{2-} + H_3O^+$$

$$HA^- + H_2O \rightleftharpoons H_2A + OH^-$$

Question 32

What is the pH of a 0.10 M NH_4Cl solution?

Answer 32

5.10

NH_4Cl is of the type BHX, the salt of a strong acid and weak base. Thus we treat this as any weak acid:

OEIOU: $pH \approx -\log \sqrt{K_a C_A}$

$= -\log \sqrt{(5.6 \times 10^{-10})(0.10)}$

$= -\log(-5.10)$

$= 5.10$

Question 33

What is the pH of a 0.10 M KF solution?

Answer 33

8.08

KF is of the type MA, the salt of a weak acid and strong base, thus a *weak base* solution:

OEIOU: $pH \approx -\log K_w / \sqrt{K_b C_B}$

(Base is F^-.)

Question 34

What is the pH of a 0.10 M NH_4F solution?

Answer 34

6.19

NH_4F is of the type BHA, the salt of a weak acid and weak base, thus an *ampholyte* solution:

OEIOU: $pH \approx -\log \sqrt{K_w K_a / K_b}$

(NH_4^+ is acid; F^- is base.)

Question 35

What is the pH of a 0.10 M $NaHCO_3$ solution?

Answer 35

8.35

$NaHCO_3$ is of the type MHA, also an *ampholyte*; HCO_3^- may act as either acid or base.

OEIOU: $pH \approx -\log \sqrt{K_w K_a / K_b}$

(Notice that there is both a K_a and a K_b for HCO_3^-.)

Question 36

What is the pH of a 0.10 M KBr solution?

Answer 36

7.00

KBr is of the type MX, the salt of a strong acid and strong base, thus gives a *neutral* solution:

OEIOU: $pH = -\log \sqrt{K_w}$

Statement 10

A **buffer solution** is a solution which resists changes in pH when moderate quantities of acid or base are added to it. One type of buffer solution is formed by mixing a **conjugate acid-base pair**, HA and A^- (or BH^+ + B, and so forth). For a buffer of this type:

$$[H^+] \approx K_a \frac{C_A}{C_B}$$

When K_a is not known, but K_b for the conjugate base is known (or vice versa), we may find the other K value from:

$$K_a K_b = K_w$$

Question 37

What is the hydrogen ion concentration in a solution formed by mixing 0.30 mole $NaHCO_3$ and 0.15 mole Na_2CO_3 in a liter of water?

Answer 37

9.6×10^{-11} M

conjugate acid-base pair:

$$HCO_3^- \text{ and } CO_3^{2-}$$

OEIOU: $[H^+] \approx K_a \dfrac{\text{moles NaHCO}_3}{\text{moles Na}_2\text{CO}_3}$

$= (4.8 \times 10^{-11} \text{ M}^2) \dfrac{0.30 \text{ mole}}{0.15 \text{ mole}}$

$= 9.6 \times 10^{-11}$ M

Question 38

What is the H^+ concentration in a solution formed by mixing 0.20 mole of HOAc and 0.10 mole NaOAc in a liter of water?

Answer 38

3.6×10^{-5} M

Conjugate acid-base pair: HOAc and OAc^-

Defining equation: $[H^+] \approx K_a \dfrac{C_A}{C_B}$

OEIOU: $[H^+] \approx K_a \dfrac{\text{moles HOAc}}{\text{moles NaOAc}}$

Question 39

What molar ratio of HOAc/NaOAc would be required to prepare a buffer solution of pH 4.50?

Answer 39

$$\frac{\text{moles HOAc}}{\text{moles NaOAc}} \approx 1.6$$

Defining equation: $[H^+] \approx K_a \dfrac{C_A}{C_B} = K_a \dfrac{\text{moles HOAc}}{\text{moles NaOAc}}$

Thus: $\dfrac{\text{moles HOAc}}{\text{moles NaOAc}} \approx \dfrac{[H^+]}{K_a}$

OEIOU: $\dfrac{\text{moles HOAc}}{\text{moles NaOAc}} \approx \dfrac{\text{antilog}(-pH)}{K_a}$

$$= \frac{\text{antilog}(-4.50)}{1.8 \times 10^{-5}}$$

$$= \frac{3.2 \times 10^{-5}}{1.8 \times 10^{-5}}$$

$$= 1.6$$

STOICHIOMETRIC PROBLEMS

Statement 11

Reactions between acids and bases generally form the foundation for **volumetric analysis** of acidic or basic constituents of "unknowns." A standard acid solution with a precisely known concentration is used to titrate a sample containing a base present in unknown quantity; or, conversely, a standard base may be used to titrate an unknown acid.

From the stoichiometry of the specific reaction, then, the concentration or percentage of the unknown can be calculated.

Question 40

The molarity of a solution of HCl is determined by titration with standard 0.100 M NaOH. If 40.00 ml of NaOH is required to titrate 50.00 ml of HCl, what is the molarity of the acid solution?

Answer 40

0.0800 M

Reaction: H^+ + OH^- → H_2O
  from HCl   from NaOH

SS: # moles HCl = # moles H⁺ = # moles OH⁻
$$= \text{\# moles NaOH}$$

Substituting: $(M \times L)_{HCl} = (M \times L)_{NaOH}$

OEIOU: $M_{HCl} = \dfrac{(M \times L)_{NaOH}}{L_{HCl}}$

$= \dfrac{0.100 \text{ M} \times 0.04000 \text{ L}}{0.05000 \text{ L}}$

$= 0.0800 \text{ M}$

Question 41

What is the molarity of a $Ba(OH)_2$ solution if 65.00 ml of 0.0500 M HCl is required to titrate 35.00 ml of the basic solution?

Answer 41

0.0464 M

$Ba(OH)_2 \rightarrow Ba^{2+} + 2OH^-$

$HCl \rightarrow H^+ + Cl^-$

Rx: $H^+ + OH^- \rightarrow H_2O$

SS: # moles $Ba(OH)_2 = \dfrac{1}{2} \times$ # moles HCl

$(M \times L)_{Ba(OH)_2} = \dfrac{1}{2}(M \times L)_{HCl}$

OEIOU: $M_{Ba(OH)_2} = \dfrac{\frac{1}{2}(M \times L)_{HCl}}{L_{Ba(OH)_2}}$

Question 42

A certain soluble solid contains CO_3^{2-} ion as the only basic species. If 0.600 g of the solid is dissolved, 32.44 ml of 0.100 M HCl is required for complete titration of the basic solution. What is the percentage of CO_3 in the solid?

Answer 42

16.2%

Rx: $2H^+ + CO_3^{2-} \rightarrow H_2CO_3$

SS: # moles $CO_3 = \frac{1}{2} \times$ # moles HCl

$\dfrac{\text{\#g } CO_3}{GMW_{CO_3}} = \dfrac{1}{2} (M \times L)_{HCl}$

OEIOU: % $CO_3 = \dfrac{1}{2} (M \times L)_{HCl} \times \dfrac{GMW_{CO_3}}{\text{g solid}} \times 100$

Question 43

A certain soluble powder contains an unknown quantity of sodium bicarbonate, $NaHCO_3$. If 25.00 ml of 0.100 N H_2SO_4 is required to titrate 0.420 g of the powder, what is the percentage of $NaHCO_3$ in the powder?

Answer 43

50.0% $NaHCO_3$

$$H_2SO_4 \rightarrow 2H^+ + SO_4^{2-}$$

$$NaHCO_3 \rightarrow Na^+ + HCO_3^-$$

Rx: $H^+ + HCO_3^- \rightarrow H_2CO_3$

SS: # moles $NaHCO_3 = 2 \times$ # moles H_2SO_4

$\dfrac{\text{\#g } NaHCO_3}{GMW_{NaHCO_3}} = 2(M \times L)_{H_2SO_4}$

$\dfrac{\text{\#g } NaHCO_3}{GMW_{NaHCO_3}} = 2\left(\dfrac{N}{n_c} \times L\right)_{H_2SO_4}$

OEIOU: % $NaHCO_3 = 2\left(\dfrac{N}{n_c} \times L\right)_{H_2SO_4} \times \dfrac{GMW_{NaHCO_3}}{\text{g powder}} \times 100$

(Recall the relationship between molarity and normality: $N = n_c M$. For H_2SO_4 $n_c = 2$.)

TESTING YOUR MASTERY—Part 17 Acids and Bases

Question 44

What is the H⁺ concentration in a solution having:

a. a 2.0×10^{-8} M OH⁻ concentration?

b. a pH of 6.301?

> **Answer 44**
>
> a. 5.0×10^{-7} M
>
> OEIOU: $[H^+] = \dfrac{K_w}{[OH^-]}$
>
> b. 5.0×10^{-7} M
>
> OEIOU: $[H^+] = \text{antilog}(-pH)$

Question 45

What is the OH⁻ concentration in a solution having:

a. a 7.8×10^{-4} M H⁺ concentration?

b. a pH of 8.114?

> **Answer 45**
>
> a. 1.3×10^{-6} M
>
> OEIOU: $[OH^-] = \dfrac{K_w}{[H^+]}$
>
> b. 1.3×10^{-6} M
>
> OEIOU: $[OH^-] = \text{antilog}(pH - 14.00)$

Question 46

What is the pH of a solution:

a. having a 0.065 M H⁺ concentration?

b. having a 0.065 M OH⁻ concentration?

Answer 46

a. 1.19

OEIOU: $pH = -\log [H^+]$

b. 12.82

OEIOU: $pH = -\log \dfrac{K_w}{[OH^-]}$

Question 47

In a solution having a pH of 3.70, what is:

a. the $[H^+]$ concentration?

b. the $[OH^-]$ concentration?

Answer 47

a. 2.0×10^{-4} M

OEIOU: $[H^+] = \text{antilog}(-pH)$

b. 5.0×10^{-11} M

OEIOU: $[OH^-] = \dfrac{K_w}{\text{antilog}(-pH)}$

Question 48

What is the H^+ concentration in:

a. a 0.086 M HBr solution?

b. a 0.086 M KOH solution?

Answer 48

a. 0.086 M

OEIOU: $[H^+] \approx C_{HBr}$

b. 1.2×10^{-13} M

OEIOU: $[H^+] \approx \dfrac{K_w}{C_{KOH}}$

Question 49

What is the H^+ concentration in a solution formed by dissolving 12.0 of HCl in 200.0 ml of water?

Answer 49

1.65 M

OEIOU: $[H^+] \approx \dfrac{(g/GMW)_{HCl}}{\#L_{solution}}$

Question 50

What is the OH^- concentration in a solution formed by dissolving 20.0 g of NaOH in 575 ml of water?

Answer 50

0.870 M

OEIOU: $[OH^-] \approx \dfrac{(g/GMW)_{NaOH}}{\#L_{solution}}$

Question 51

What is the OH^- concentration in:

a. a 0.020 M LiOH solution?

b. a 0.020 M $Ca(OH)_2$ solution?

Answer 51

a. 0.020 M

OEIOU: $[OH^-] \approx C_{LiOH}$

b. 0.040 M

OEIOU: $[OH^-] \approx 2C_{Ca(OH)_2}$

Question 52

What is the H^+ concentration in a 0.33 M solution of:

a. HI?

b. HF?

Answer 52

a. 0.33 M

OEIOU: $[H^+] \approx C_{HI}$ (strong acid)

b. 1.5×10^{-2} M

OEIOU: $[H^+] \approx \sqrt{K_a C_A}$ (weak acid)

Question 53

In a 0.25 M HCN solution, what is the approximate concentration of:

a. H^+?

b. OH^-?

Answer 53

a. 1.3×10^{-5} M

OEIOU: $[H^+] \approx \sqrt{K_a C_A}$

b. 7.7×10^{-10} M

OEIOU: $[OH^-] \approx \dfrac{K_w}{\sqrt{K_a C_A}}$

Question 54

In a 0.25 M NaCN solution, what is the approximate concentration of:

a. H^+?

b. OH^-?

Answer 54

a. 4.0×10^{-12} M

OEIOU: $[H^+] \approx \dfrac{K_w}{\sqrt{K_b C_B}}$

b. 2.5×10^{-3} M

OEIOU: $[OH^-] \approx \sqrt{K_b C_B}$

Question 55

What is the OH⁻ concentration in a 0.40 M solution of:

a. KOH?

b. KCN?

Answer 55

a. 0.40 M

OEIOU: $[OH^-] \approx C_{KOH}$ (strong base)

b. 3.2×10^{-3} M

OEIOU: $[OH^-] \approx \sqrt{K_b C_B}$ (weak base)

Question 56

What is the H⁺ concentration in a 0.28 M solution of:

a. NH_4Br?

b. NH_3?

Answer 56

a. 1.2×10^{-5} M

OEIOU: $[H^+] \approx \sqrt{K_a C_A}$

b. 4.4×10^{-12} M

OEIOU: $[OH^-] \approx \dfrac{K_w}{\sqrt{K_b C_B}}$

Question 57

What is the OH⁻ concentration in a 0.30 M solution of:

a. Na_2CO_3?

b. $NaHCO_3$?

Answer 57

a. 7.9×10^{-3} M

OEIOU: $[OH^-] \approx \sqrt{K_{b_1} C_B}$

b. 2.2×10^{-6} M

OEIOU: $[OH^-] \approx \dfrac{K_w}{\sqrt{K_a K_w / K_b}}$

Question 58

In a 0.15 M H_2S solution, what is the molar concentration of:

a. H^+?
b. SH^-?
c. S^{2-}?

Answer 58

a. 1.3×10^{-4} M

OEIOU: $[H^+] \approx \sqrt{K_{a_1} C_A}$

b. 1.3×10^{-4} M

OEIOU: $[SH^-] \approx \sqrt{K_{a_1} C_A}$

c. 1.0×10^{-14} M

OEIOU: $[S^{2-}] \approx K_{a_2}$

Question 59

What is the H^+ concentration in a solution of HF which is:

a. 0.58 M?
b. 0.058 M?

Answer 59

a. 2.1×10^{-2} M

OEIOU: $[H^+] \approx \sqrt{K_a C_A}$

b. 6.3×10^{-3} M

OEIOU: $K_a \approx \dfrac{[H^+]^2}{C_A - [H^+]}$

Solve quadratic for $[H^+]$.

Note that the usual approximation cannot be made in (b) because $\sqrt{K_a C_A} = 6.4 \times 10^{-3}$ which is greater than $0.05\, C_A = 2.9 \times 10^{-3}$.

Question 60

What is the pH of a 0.10 M solution of:

a. NaCl?

b. NaF?

Answer 60

a. 7.00 (neutral)

OEIOU: $\text{pH} = -\log \sqrt{K_w}$

(salt of strong acid and strong base)

b. 8.08 (basic)

OEIOU: $\text{pH} \approx -\log \dfrac{K_w}{\sqrt{K_b C_B}}$

(salt of weak acid and strong base)

Question 61

What is the pH of a 0.10 M solution of:

a. KI?

b. NH$_4$I?

Answer 61

a. 7.00 (neutral)

OEIOU: $\text{pH} = -\log \sqrt{K_w}$

(salt of strong acid and strong base)

b. 5.12 (acidic)

OEIOU: $\text{pH} \approx -\log \sqrt{K_a C_A}$

(salt of strong acid and weak base)

Question 62

What is the pH of a 0.16 M $KHCO_3$ solution?

Answer 62

8.35

OEIOU: $pH \approx -\log\sqrt{\dfrac{K_w K_a}{K_b}}$

Question 63

Calculate the approximate concentrations of all molecular and ionic species (except H_2O) in a 0.20 M solution of:

a. HNO_3

b. LiOH

Answer 63

a. $[H^+] \approx 0.20$ M

$[NO_3^-] \approx 0.20$ M

$[OH^-] \approx 5.0 \times 10^{-14}$ M

OEIOU: $[H^+] \approx [NO_3^-] = C_{HX}$

OEIOU: $[H^+] \approx K_w/C_{HX}$

b. $[OH^-] \approx 0.20$ M

$[Li^+] \approx 0.20$ M

$[H^+] \approx 5.0 \times 10^{-14}$ M

OEIOU: $[OH^-] \approx [Li^+] = C_{MOH}$

OEIOU: $[H^+] \approx K_w/C_{MOH}$

Question 64

Calculate the approximate concentrations of all molecular and ionic species (except H_2O) in a 0.20 M solution of HCN.

Answer 64

$[H^+] \approx 1.2 \times 10^{-5}$ M

$[CN^-] \approx 1.2 \times 10^{-5}$ M

$[HCN] \approx 0.20$ M

$[OH^-] \approx 8.3 \times 10^{-10}$ M

OEIOU: $[H^+] \approx [CN^-] \approx \sqrt{K_a C_A}$

OEIOU: $[HCN] \approx C_A$

OEIOU: $[OH^-] \approx K_w/\sqrt{K_a C_A}$

Question 65

Calculate the approximate concentrations of all molecular and ionic species (except H_2O) in a 0.20 M solution of KOAc.

Answer 65

$[OH^-] \approx 1.1 \times 10^{-5}$ M

$[HOAc] \approx 1.1 \times 10^{-5}$ M

$[OAc^-] \approx 0.20$ M

$[K^+] = 0.20$ M

$[H^+] \approx 9.1 \times 10^{-10}$ M

OEIOU: $[OH^-] \approx [HOAc] \approx \sqrt{K_b C_B}$

OEIOU: $[OAc^-] \approx [K^+] = C_B$

OEIOU: $[H^+] \approx K_w/\sqrt{K_b C_B}$

Question 66

Calculate the approximate concentrations of all molecular and ionic species (except H_2O) in a 0.20 M solution of H_2S.

Answer 66

$[H^+] \approx 1.5 \times 10^{-4}$ M

$[HS^-] \approx 1.5 \times 10^{-4}$ M

$[H_2S] \approx 0.20$ M

$[S^{2-}] \approx 1.0 \times 10^{-14}$ M

$[OH^-] \approx 6.7 \times 10^{-11}$ M

OEIOU: $[H^+] \approx [HS^-] \approx \sqrt{K_{a,1}C_A}$

OEIOU: $[H_2S] \approx C_A$

OEIOU: $[S^{2-}] \approx K_{a,2}$

OEIOU: $[OH^-] \approx K_w/\sqrt{K_{a,1}C_A}$

Question 67

Calculate the approximate concentrations of all molecular and ionic species (except H_2O) in a 0.20 M solution of Na_2CO_3.

Answer 67

$[OH^-] \approx 6.5 \times 10^{-3}$ M

$[HCO_3^-] \approx 6.5 \times 10^{-3}$ M

$[CO_3^{2-}] \approx 0.20$ M

$[H_2CO_3] \approx 2.4 \times 10^{-8}$ M

$[H^+] \approx 1.5 \times 10^{-12}$ M

$[Na^+] = 0.40$ M

OEIOU: $[OH^-] \approx [HCO_3^-] \approx \sqrt{K_{b,1}C_B}$

OEIOU: $[CO_3^{2-}] \approx C_B$

OEIOU: $[H_2CO_3] \approx K_{b,2}$

OEIOU: $[H^+] \approx K_w/\sqrt{K_{b,1}C_B}$

OEIOU: $[Na^+] = 2C_B$

Question 68

The OH^- concentration in a 0.20 M solution of an unknown weak base is 2.5×10^{-5} M. Find K_b for the base.

Answer 68

3.1×10^{-9}

Question 69

What is the H^+ concentration in a solution formed by mixing 0.20 mole of NH_4Cl and 0.45 mole of NH_3 in 500.0 ml of water?

Answer 69

2.5×10^{-10} M

OEIOU: $[H^+] \approx K_a \dfrac{\text{moles } NH_4Cl}{\text{moles } NH_3}$

Question 70

What molar ratio of $NaHCO_3/Na_2CO_3$ is needed to give a buffer solution of pH 9.15?

Answer 70

$\dfrac{NaHCO_3}{Na_2CO_3} \approx 15$

OEIOU: $\dfrac{\text{moles } NaHCO_3}{\text{moles } Na_2CO_3} \approx \dfrac{\text{antilog}(-pH)}{K_a}$

Question 71

When 0.48 mole of HF and 0.33 mole of NaF are mixed in 750 ml of water, the pH is 2.99. What will be the pH after the addition of:

a. 0.02 mole of HI?

b. 0.02 mole of KOH?

Answer 71

a. 2.95

Reaction: $H^+ + F^- \rightarrow HF$

0.02 mole used up   0.02 mole formed

OEIOU: $pH \approx -\log K_a \dfrac{\text{moles HF}}{\text{moles NaF}}$

b. 3.04

Reaction: $OH^- + HF \rightarrow F^- + H_2O$

0.02 mole used up   0.02 mole formed

OEIOU: $pH \approx -\log K_a \dfrac{\text{moles HF}}{\text{moles NaF}}$

Question 72

A solution of HI is titrated with 0.113 M KOH. If 60.13 ml of KOH is required to titrate 46.75 ml of HI, what is the molarity of the HI?

Answer 72

0.145 M

OEIOU: $M_{HI} = \dfrac{(M \times L)_{KOH}}{L_{HI}}$

Question 73

Soda ash (impure Na_2CO_3) is analyzed by titrating with 0.200 M HCl. If 0.740-g sample of soda ash requires 60.10 ml of acid to titrate the carbonate completely, what is the percentage of Na_2CO_3 in the soda ash?

Answer 73

86.1% Na_2CO_3

OEIOU: $\% \, Na_2CO_3 = \dfrac{1}{2}(M \times L)_{HCl} \times \dfrac{GMW_{Na_2CO_3}}{\text{g sample}} \times 100$

PART 18
OXIDATION-REDUCTION PROCESSES

Statement 1

For a strip of metal, M, dipping into a solution of its ions, M^{n+}, we may write the following equilibrium expression:

$$M^{n+} + ne^- \rightleftharpoons M$$

where e^- represents the electron. Under these conditions an insignificant number of M atoms will give up ne^- to form M^{n+} ions, nor will there be extra electrons in the solution to reduce M^{n+} to M.

If, however, a second metal, M', is dipped into a solution of its ions, M'^{m+}, and if the two metals, M and M', are joined by an external wire while the two solutions containing M^{n+} and M'^{m+} are connected by a "salt bridge" or porous barrier, electrons may be supplied from one metal to shift the equilibrium involving the other. An apparatus in which this electron exchange takes place spontaneously is called a **galvanic cell**.

We may view this exchange as a competition for electrons between M^{n+} and M'^{m+}, with the ion having the greatest affinity for electrons "winning" and becoming **reduced** to metal atoms. One measure of an ion's (or other species') relative affinity for electrons is its **standard reduction potential**, ϵ°.

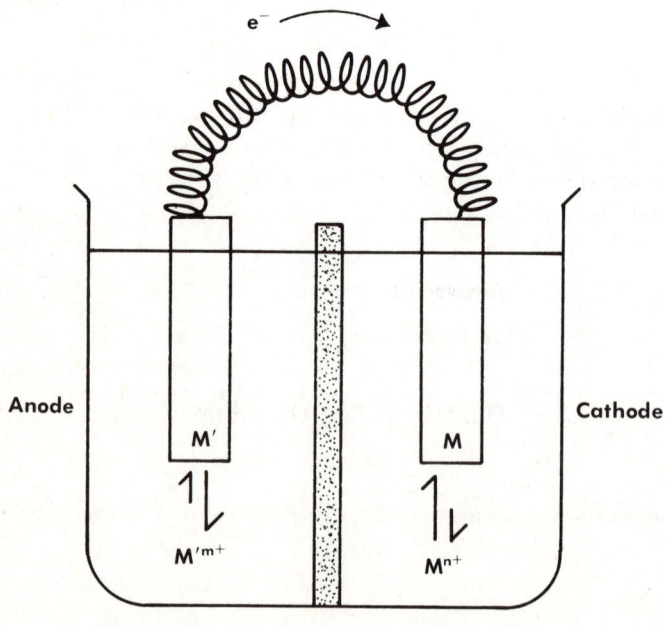

A galvanic cell in which $\epsilon^\circ_{M^{n+}} > \epsilon^\circ_{M'^{m+}}$

TABLE 18.1 STANDARD REDUCTION POTENTIALS (volts)

K^+	$\rightarrow K$	$-2.93v$	Sn^{2+}	$\rightarrow Sn$	$-0.14v$	Ag^+	$\rightarrow Ag$	$+0.80v$
Ca^{2+}	$\rightarrow Ca$	$-2.87v$	Pb^{2+}	$\rightarrow Pb$	$-0.13v$	Br_2	$\rightarrow Br$	$+1.06v$
Mg^{2+}	$\rightarrow Mg$	$-2.37v$	$2H^+$	$\rightarrow H_2$	$0.00v$	O_2	$\rightarrow H_2O$	$+1.23v$
Al^{3+}	$\rightarrow Al$	$-1.66v$	SO_4^{2-}	$\rightarrow SO_2$	$+0.20v$	$Cr_2O_7^{2-}$	$\rightarrow 2Cr^{3+}$	$+1.33v$
H_2O	$\rightarrow H_2$	$-0.83v$	Cu^{2+}	$\rightarrow Cu$	$+0.34v$	Cl_2	$\rightarrow 2Cl^-$	$+1.36v$
Zn^{2+}	$\rightarrow Zn$	$-0.76v$	O_2	$\rightarrow OH^-$	$+0.40v$	MnO_4^-	$\rightarrow Mn^{2+}$	$+1.52v$
Cd^{2+}	$\rightarrow Cd$	$-0.40v$	I_2	$\rightarrow 2I^-$	$+0.54v$	$S_2O_8^{2-}$	$\rightarrow SO_4^{2-}$	$+2.01v$
$PbSO_4$	$\rightarrow Pb$	$-0.36v$	Fe^{3+}	$\rightarrow Fe^{2+}$	$+0.77v$	F_2	$\rightarrow F^-$	$+2.87v$
Ni^{2+}	$\rightarrow Ni$	$-0.25v$						

The **more positive** ϵ°, the greater the likelihood that the reduction will take place.

ϵ° values are determined at unit activity. For simplicity we will assume that this is about the same as one molar for dissolved species; activities of solids in contact with a solution are taken as being unity. ϵ° values for a variety of species can be found in Table 18.1.

Question 1
Would Cu^{2+}, or Zn^{2+}, have a greater affinity for electrons?

Answer 1
Cu^{2+}

$$\epsilon^\circ_{Cu} = +0.34v > -0.76v = \epsilon^\circ_{Zn}$$

Question 2
A Cu strip is dipped into a 1.00 M $CuSO_4$ solution in one side of a cell and a Zn strip is dipped into a 1.00 M $ZnSO_4$ solution in the other side. Which reduction process will "win out" when the metal strips are connected by an external wire?

Answer 2
$$Cu^{2+} + 2e^- \rightleftharpoons Cu$$

Because the $\epsilon^\circ_{Cu} > \epsilon^\circ_{Zn}$, electrons will flow from the Zn side to the Cu side, supplying electrons to the Cu^{2+} ions.

Question 3

If Cd and Ni are dipped into 1.00 M solutions of $Cd(NO_3)$ and $Ni(NO_3)_2$, respectively, in a cell, which reduction process will take place when the metals are joined by an external wire?

Answer 3

$$Ni^{2+} \; 2e^- \rightleftharpoons Ni$$

$\epsilon^{\circ}_{Ni} > \epsilon^{\circ}_{Cd}$, so Ni^{2+} will "win out" in the competition for e^-.

Statement 2

While **reduction** is taking place in one half of the cell, called the **cathode**, **oxidation** (loss of electrons) is taking place in the other side (the **anode**):

$$\text{reduction} \quad M^{n+} + ne^- \rightleftharpoons M \quad \text{cathode}$$

$$\text{oxidation} \quad M' \rightleftharpoons M'^{m+} + me^- \quad \text{anode}$$

The number of electrons gained must equal the number lost, so considering the overall **cell reaction**:

$$mM^{n+} + nM' \rightleftharpoons nM'^{n+} + mM$$

The **cell diagram** provides a shorthand way of representing the cell:

$$M' \,|\, M'^{m+} \,||\, M^{n+} \,|\, M$$
$$\text{anode} \qquad\quad \text{cathode}$$

where the single vertical solid lines represent the separation of the solid metal electrode from the solution, and the double lines represent the porous barrier or salt bridge connecting the anode and cathode chambers.

Question 4

Write the cell diagram for the Cu-Zn cell in Question 2.

Answer 4

$$Zn \,|\, Zn^{2+} \,||\, Cu^{2+} \,|\, Cu$$
$$\text{anode} \qquad\quad \text{cathode}$$

Question 5

Write the cell diagram for the Ni-Cd cell in Question 3.

Answer 5

Cd | Cd^{2+} || Ni^{2+} | Ni
anode   cathode

Question 6

Write the cell reaction for the following cell:

Al | Al^{3+} || Ag$^+$ | Ag

Answer 6

Al + 3Ag$^+$ ⇌ 3Ag + Al^{3+}

Question 7

Write the cell reaction for this cell:

Cr | Cr^{3+} || Pb^{2+} | Pb

Answer 7

2Cr + 3Pb^{2+} ⇌ 3Pb + 2Cr^{3+}

Statement 3

The same oxidation-reduction process taking place in a galvanic cell will take place by the **direct mixing** of the **oxidized form** of one species with the **reduced** form of another. We may, in fact, predict the **"driving force"** of an oxidation-reduction reaction by using **half-cell potentials** to calculate the **cell potential** for the theoretical cell in which the same net reaction would take place:

$$\epsilon^{\circ}_{cell} = \epsilon^{\circ}_{cathode} - \epsilon^{\circ}_{anode}$$

If ϵ°_{cell} is **positive**, the reaction is spontaneous, and the more positive ϵ°_{cell} is, the greater the driving force of the reaction.

If ϵ°_{cell} is **negative**, the reaction will not take place spontaneously as written (the opposite reaction will take place).

Note: Not all chemists follow the same conventions in treating half-cell and cell potentials. Oxidation potentials are often used rather than reduction potentials. These have the same absolute value but opposite signs, and $\epsilon°_{cell} = \epsilon°_{anode} - \epsilon°_{cathode}$. A third convention used oxidation potentials for anodes and reduction potentials for cathodes, and $\epsilon°_{cell} = \epsilon°_{cathode} + \epsilon°_{anode}$. The same value for cell potential is obtained by either convention.

Question 8

Calculate the cell potential of the Cu-Zn cell in Question 2.

Answer 8

+1.10v

$$\epsilon°_{cell} = \epsilon°_{cathode} - \epsilon°_{anode}$$

$$= \epsilon°_{Cu} - \epsilon°_{Zn}$$

$$= +0.34v - (-0.76v)$$

$$= +1.10v$$

Question 9

Tin metal is placed in a 1.00 M $AgNO_3$ solution. Will the Sn reduce the Ag^+ to give Ag metal?

Answer 9

Yes.

Theoretical cell: $Sn \mid Sn^{2+} \parallel Ag^+ \mid Ag$

$$\epsilon°_{cell} = \epsilon°_{cathode} - \epsilon°_{anode}$$

$$= \epsilon°_{Ag} - \epsilon°_{Sn}$$

$$= +0.80v - (-0.14v)$$

$$= +0.94v$$

$\epsilon°_{cell}$ is +, so Sn *will* reduce Ag^+.

Question 10

If Cu metal is placed in a 1 M hydrochloric acid solution, will the Cu "displace" the hydrogen; that is, will the following reaction take place:

$$Cu + 2H^+ \rightleftharpoons H_2 \uparrow + Cu^{2+}$$

Answer 10

No.

Theoretical cell: $Cu \mid Cu^{2+} \parallel H^+ \mid H_2, Pt$

$$\epsilon°_{cell} = \epsilon°_{H_2} - \epsilon°_{Cu}$$
$$= 0.00v - (+0.34v)$$
$$= -0.34v$$

$\epsilon°_{cell}$ is $-$, so Cu *will not* reduce H^+.

(Note that when one species, such as H_2, is not a solid and cannot actually be used for an electrode, an inert metal, Pt, is shown as the electrode.)

Statement 4

Balancing oxidation-reduction equations is often more complicated than balancing acid-base equations or precipitation equations. The **ion-electron method** is a relatively simple approach to balancing these equations. One looks first at individual half-cell reactions, then combines these to get the overall balanced equation.

Rule	Example
	$Cr_2O_7^{2-} + H_2SO_3 \rightleftharpoons Cr^{3+} + SO_4^{2-}$ (acid)
(1) Write half-reactions separately, with **atoms balanced** except for O and H	(1) $Cr_2O_7^{2-} \rightleftharpoons 2Cr^{3+}$ $H_2SO_3 \rightleftharpoons SO_4^{2-}$
(2) If there are **extra O** on one side, balance these as follows, depending on whether the solution is acid or basic:	(2) $14H^+ + Cr_2O_7^{2-} \rightleftharpoons 2Cr^{3+} + 7H_2O$ $H_2O + H_2SO_3 \rightleftharpoons SO_4^{2-} + 2H^+$

$2H^+ + (\text{extra O}) \rightleftharpoons H_2O$
(acid solution)

$H_2O + (\text{extra O}) \rightleftharpoons 2OH^-$
(basic solution)

(3) If there are **extra H** on one side, balance these as follows:

(3) $14H^+ + Cr_2O_7^{2-} \rightleftharpoons 2Cr^{3+} + 7H_2O$

$H_2O + H_2SO_3 \rightleftharpoons SO_4^{2-} + 4H^+$

$(\text{extra H}) \rightleftharpoons H^+$
(acid solution)

$OH^- + (\text{extra H}) \rightleftharpoons H_2O$
(basic solution)

(4) Balance charges by adding electrons to the side with more positive charge.

(4) $6e^- + 14H^+ + Cr_2O_7^{2-} \rightleftharpoons 2Cr^{3+} + 7H_2O$

$H_2SO_3 + H_2O \rightleftharpoons SO_4^{2-} + 4H^+ + 2e^-$

(5) Balance electron transfer, so that the same number of electrons are gained in one half-cell reaction as are lost in the other, multiplying half-reactions by appropriate coefficients.

(5) $6e^- + 14H^+ + Cr_2O_7^{2-} \rightleftharpoons 2Cr^{3+} + 7H_2O$

$3(H_2SO_3 + H_2O \rightleftharpoons SO_4^{2-} + 4H^+ + 2e^-)$

(6) Add the half-reactions.

(6) $2H^+ + Cr_2O_7^{2-} + 3H_2SO_3 \rightleftharpoons 2Cr^{3+} + 4H_2O + 3SO_4^{2-}$

(7) Check for atom and charge balance.

(7)

	left	right	
Cr	2	2	
O	16	16	
H	8	8	
S	3	3	
charge	0	0	balanced!

Question 11

Complete and balance the following equations:

a. $Fe^{3+} + I^- \rightleftharpoons I_2 + Fe^{2+}$

b. $Fe^{2+} + Cr_2O_7^{2-} \rightleftharpoons Cr^{3+} + Fe^{3+}$
(acid solution)

Answer 11

a. $2Fe^{3+} + 2I^- \rightleftharpoons I_2 + 2Fe^{2+}$

b. $6Fe^{2+} + Cr_2O_7^{2-} + 14H^+ \rightleftharpoons 2Cr^{3+} + 7H_2O + 6Fe^{3+}$

Question 12

Complete and balance the following equations:

a. $Al + H^+ \rightleftharpoons H_2 + Al^{3+}$

b. $BrO_3^- + I^- \rightleftharpoons I_2 + Br^-$ (acid solution)

Answer 12

a. $2Al + 6H^+ \rightleftharpoons 3H_2 + 2Al^{3+}$

b. $BrO_3^- + 6H^+ + 6I^- \rightleftharpoons 3I_2 + Br^- + 3H_2O$

Question 13

Complete and balance these equations:

a. $MnO_4^- + Br^- \rightleftharpoons Br_2 + Mn^{2+}$
   (acid solution)

b. $MnO_4^- + Br^- \rightleftharpoons Br_2 + MnO_2$
   (basic solution)

Answer 13

a. $2MnO_4^- + 16H^+ + 10Br^- \rightleftharpoons 5Br_2 + 2Mn^{2+} + 8H_2O$

b. $2MnO_4^- + 4H_2O + 6Br^- \rightleftharpoons 3Br_2 + 2MnO_2 + 8OH^-$

Question 14

Complete and balance the following equations:

a. $H_2S + ClO_3^- \rightleftharpoons Cl^- + S$ (acid solution)

b. $HOI \rightleftharpoons IO_3^- + I^-$ (basic solution)

Answer 14

a. $3H_2S + ClO_3^- \rightleftharpoons Cl^- + 3H_2O + 3S$

b. $3OH^- + 3HOI \rightleftharpoons IO_3^- + 2I^- + 3H_2O$

Statement 5

The availability of ϵ values for half-reactions makes it possible, as we have seen, to use ϵ^o_{cell} to predict the likelihood that a given oxidation-reduction reaction will occur. Qualitative predictions can be made, however, without actually doing calculations but merely by comparing the positions of two half-reactions in an **activity series** (or "electromotive" series); that is, comparing ϵ^o values.

An activity series is customarily constructed by listing half-reactions in order of increasing reduction potential, ϵ^o (see Table 18.1). If we write a generalized half-reaction in this form:

$$\underset{\text{oxidixed form}}{\text{Ox}} + ne^- \rightleftharpoons \underset{\text{reduced form}}{\text{Red}}$$

the **oxidized form** from one half-reaction will **oxidize** the **reduced form** from any half-reaction **above** it in the activity series. Conversely, the **reduced form** will **reduce** the **oxidized form** from any half-reaction **below** it in the series. Thus, for the overall reaction:

$$\underset{\text{oxidizing agent}}{\text{Ox}_2} + \underset{\text{reducing agent}}{\text{Red}_1} \rightleftharpoons \text{Ox}_1 + \text{Red}_2$$

the oxidized species from the lower position in the activity series should be a suitable **oxidizing agent** for the **reduced** species **above** it.

Conversely, the reduced species is a suitable **reducing agent** for the **oxidized** species **below** it.

Question 15

Will Cu^{2+} oxidize Pb metal to give Pb^{2+}?

Answer 15

Yes.

Cu^{2+} is the *oxidized* form in the half-reaction:

$$Cu^{2+} + 2e^- \rightleftharpoons Cu \quad \epsilon^o_{Cu} = +0.34v$$

which lies *below*:

$$Pb^{2+} + 2e^- \rightleftharpoons Pb \quad \epsilon^o_{Pb} = -0.13v$$

Question 16

Will I_2 oxidize Fe^{2+} to Fe^{3+}?

Answer 16

No.

I_2 is the *oxidized* form in the half-reaction:

$$I_2 + 2e^- \rightleftharpoons 2I^- \qquad \epsilon^\circ_{I^-} = +0.54v$$

which lies *above*:

$$Fe^{3+} + e^- \rightleftharpoons Fe^{2+} \qquad \epsilon^\circ_{Fe} = +0.77v$$

Question 17

Will Al metal reduce H^+ to H_2 gas?

Answer 17

Yes.

Al is the *reduced* form in the half-reaction:

$$Al^{3+} + 3e^- \rightleftharpoons Al \qquad \epsilon^\circ_{Al} = -1.66$$

which lies *above*:

$$2H^+ + 2e^- \rightleftharpoons H_2 \qquad \epsilon^\circ_{H_2} = 0.00v$$

Question 18

Arrange these metals in order of decreasing reducing strength:

K, Zn, Mg, Ca

Answer 18

K > Ca > Mg > Zn

This is the order in which these metals are listed as reduced forms in the activity series.

Question 19

Arrange these species in order of decreasing oxidizing strength in acid solution:

Ag^+, MnO_4^-, $Cr_2O_7^{2-}$, Cl_2

Answer 19

$$MnO_4^- > Cl_2 > Cr_2O_7^{2-} > Ag^+$$

This is the *reverse* order in which these are listed as oxidized forms in the activity series.

Question 20

Which of these metals will "displace" H^+ from H_2SO_4 to form H_2 gas?

Answer 20

Al and Ni

Al and Ni lie above H^+ in the activity series.

Ag lies below H^+ in the activity series.

Question 21

In which of these solutions will Zn metal dissolve to give Zn^{2+}?

$Fe(NO_3)_2$, $Mg(NO_3)_2$, $Pb(NO_3)_2$

Answer 21

$Fe(NO_3)_2$ and $Pb(NO_3)_2$

Question 22

On which of these metals should Cu^{2+} "plate out" as Cu metal when a strip of the metal in question is dipped into $Cu(NO_3)_2$ solution?

Ag, Pb, Zn

Answer 22

Pb and Zn

Statement 6

Predictions based on an activity series or on ϵ_{cell}° values are totally reliable only when the concentrations of dissolved species involved in the reaction are unity (actually, when the activity is unity). Slight differences between ϵ_{cell}° and ϵ_{cell}, and thus in the driving force of the oxidation-reduction reaction, are observed **when concentrations are significantly different from unity**.

The **Nernst equation** can be used to take into account **concentration effects** in oxidation-reduction reactions. At 25°C, for the general reaction:

$$aA + bB \rightleftharpoons cC + dD$$

$$\epsilon_{cell} = \epsilon_{cell}^{\circ} - \frac{0.0591}{n} \log \frac{[C]_i^c [D]_i^d}{[A]_i^a [B]_i^b}$$

where n represents the number of electrons exchanged per equation and the concentrations are initial concentrations. Water and solid reactants or products may be assumed to have unit activity when using this equation.

As with ϵ_{cell}°, a **positive** ϵ_{cell} value indicates that a reaction will take place as written.

Question 23

Write the Nernst equation for these reactions:

a. $Fe^{2+} + Ce^{4+} \rightleftharpoons Ce^{3+} + Fe^{3+}$

b. $5Fe^{2+} + MnO_4^- + 8H^+ \rightleftharpoons 5Fe^{3+} + Mn^{2+} + 4H_2O$

Answer 23

a. $\epsilon_{cell} = \epsilon_{cell}^{\circ} - \frac{0.0591}{n} \log \frac{[Fe^{3+}][Ce^{3+}]}{[Fe^{2+}][Ce^{4+}]}$

b. $\epsilon_{cell} = \epsilon_{cell}^{\circ} - \frac{0.0591}{n} \log \frac{[Fe^{3+}]^5 [Mn^{2+}]}{[Fe^{2+}]^5 [MnO_4^-][H^+]^8}$

Question 24

Calculate ϵ_{cell} for the following cell at 25°C:

$$Zn \mid Zn\ (0.0800\ M) \parallel Cu^{2+}\ (0.0400\ M) \mid Cu$$

Answer 24

1.09v

Rx: $Zn + Cu^{2+} \rightleftharpoons Cu + Zn^{2+}$

$$\epsilon_{cell} = \epsilon^{o}_{cell} - \frac{0.0591}{n} \log \frac{[Zn^{2+}]}{[Cu^{2+}]}$$

OEIOU: $\epsilon_{cell} = \epsilon^{o}_{Cu} - \epsilon^{o}_{Zn} - \frac{0.0591}{n} \log \frac{[Zn^{2+}]}{[Cu^{2+}]}$

$$= 0.34v - (-0.76v) - \frac{0.0591}{2} \log \frac{(0.0800)}{(0.0400)}$$

$$= 1.10v - 0.009v$$

$$= 1.09v$$

Question 25

Will the following reaction take place when Pb powder is stirred into a solution which is 2.00 M in SO_4^{2-} and 1.50 M in Cd^{2+}?

$$Pb + Cd^{2+} + SO_4^{2-} \rightleftharpoons PbSO_4 + Cd$$

Answer 25

No.

OEIOU: $\epsilon_{cell} = \epsilon^{o}_{Cd} - \epsilon^{o}_{PbSO_4} - \frac{0.0591}{n} \log \frac{1}{[Cd^{2+}][SO_4^{2-}]}$

$$= -0.40v - (-0.36v) - \frac{0.0591}{2} \log \frac{1}{(1.50)(2.00)}$$

$$= -0.04v + 0.014v$$

$$= -0.03v$$

ϵ_{cell} is slightly *negative*, so no reaction is expected.

Statement 7

In addition to predictions as to **whether** a given oxidation-reduction reaction will take place, we may predict **how far** the reaction will proceed. For the general reaction:

$$aA + bB \rightleftharpoons cC + dD$$

we may write an **equilibrium constant**:

$$K_e = \frac{[C]^c [D]^d}{[A]^a [B]^b}$$

which at 25°C is related to ϵ_{cell}° by the following equation:

$$\epsilon_{cell}^{\circ} = \frac{0.0591}{n} \log K_e$$

Knowing the value of K_e, we can predict the following:

if K_e is very **large**, the reaction goes essentially **to completion**;

if K_e is very **small**, there is essentially **no reaction**;

if K_e is **intermediate** in value, there is **partial reaction**, with significant quantities of reactants and products mixed at equilibrium.

Question 26

ϵ_{cell}° for the reaction $Cu^{2+} + Zn \rightleftharpoons Zn^{2+} + Cu$ is 1.10v.

a. What is K_e at 25°C for this reaction?

b. Does the reaction go to completion?

Answer 26

a. 1.6×10^{37}

b. Yes. (Large K_e)

OEIOU: $K_e = \text{antilog} \left[\frac{n \, \epsilon_{cell}^{\circ}}{0.0591} \right]$

Question 27

What is the molar ratio of Cu^{2+} to Zn^{2+} at equilibrium when Zn is dissolved in $CuSO_4$ solution?

Answer 27

6.3×10^{-38}

$$\epsilon_{cell}^{\circ} = \frac{0.0591}{n} \log \frac{[Zn^{2+}]}{[Cu^{2+}]}$$

$$= - \frac{0.0591}{n} \log \frac{[Cu^{2+}]}{[Zn^{2+}]}$$

OEIOU: $\dfrac{[Cu^{2+}]}{[Sn^{2+}]}$ = antilog $\left[-\dfrac{n\,\epsilon°_{cell}}{0.0591}\right]$

$= 6.3 \times 10^{-38}$

Not much Cu^{2+} left!

Question 28

Does the following reaction go essentially to completion:

$$Sn + Pb^{2+} \rightleftharpoons Pb + Sn^{2+}$$

Answer 28

No. K_e is not very large, so a significant mixture would exist at equilibrium.

$$K_e = \text{antilog}\left[\dfrac{n\,\epsilon°_{cell}}{0.0591}\right]$$

$$= 2.18$$

Question 29

What is the molar ratio of Pb^{2+} to Sn^{2+} at equilibrium when Sn metal is dissolved in $Pb(NO_3)_2$ solution?

Answer 29

0.46

In other words, there is nearly half as much unreacted Pb^{2+} as Sn^{2+}.

Statement 8

In an **electrolytic cell** a **nonspontaneous oxidation-reduction process** is brought about by imposing an electrical potential across the cell, thereby forcing current to flow through the solution or molten salt into which the electrodes are dipping.

PART 18—OXIDATION-REDUCTION PROCESSES

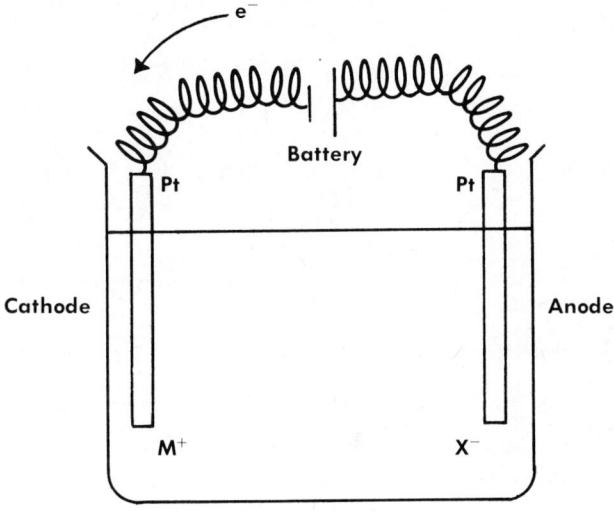

Typical electrolytic cell

Particularly in the **electrolysis of aqueous solutions**, where H_2O, H^+, and OH^- are always present in addition to ions from the salt, several possible cathode and anode reactions are possible. When inert electrodes such as Pt are used, typical processes are as follows.

Cathode Processes

a. $M^{n+} + ne^- \rightarrow M$

b. $2H^+ + 2e^- \rightarrow H_2$
   (acid solution)

c. $2H_2O + 2e^- \rightarrow H_2 + 2OH^-$
   (neutral or basic solution)

Anode Processes

a. $2X^- \rightarrow X_2 + 2e^-$

b. oxidation of oxygenated anion

c. $2H_2O \rightarrow O_2 + 4H^+ + 4e^-$
   (neutral or acid solution)

d. $4OH^- \rightarrow O_2 + 2H_2O + 4e^-$
   (basic solution)

Unless a high concentration of H^+ or OH^- is present, water electrolysis may be expected to involve the reduction or oxidation of the H_2O molecule.

Question 29a

When molten KBr is electrolyzed, what are:

a. the possible cathode processes?

b. the possible anode processes?

Answer 29a

a. $K^+ + e^- \to K$

b. $2Br^- \to Br_2 + 2e^-$

In a molten salt no H_2O, H^+, or OH^- is present, so no processes involving them need be considered.

Question 30

When an aqueous solution of KBr undergoes electrolysis, what are:

a. the possible cathode processes?

b. the possible anode processes?

Answer 30

a. $K^+ + e^- \to K$

$2H_2O + 2e^- \to H_2 + 2OH^-$

b. $2Br^- \to Br_2 + 2e^-$

$2H_2O \to O_2 + 2H^+ + 2e^-$

In this neutral aqueous salt solution we must consider the electrolysis of H_2O as a possibility.

Question 31

When an aqueous solution of $CdCl_2$ dissolved in 1 M HCl undergoes electrolysis, what are:

a. the possible cathode processes?

b. the possible anode processes?

Answer 31

a. $Cd^{2+} + 2e^- \to Cd$

$2H^+ + 2e^- \to H_2$

b. $2Cl^- \to Cl_2 + 2e^-$

$2H_2O \to O_2 + 2H^+ \; 2e^-$

PART 18–OXIDATION-REDUCTION PROCESSES 377

This is an *acid* solution, so there is a high concentration of H^+. As a consequence, H^+ rather than H_2O should be considered in the cathode processes.

Question 32

When an aqueous solution of $Ba(OH)_2$ undergoes electrolysis, what are:

a. the possible cathode processes?

b. the possible anode processes?

Answer 32

a. $Ba^{2+} + 2e^- \rightarrow Ba$

$2H_2O + 2e^- \rightarrow H_2 + 2OH^-$

b. $4OH^- \rightarrow O_2 + 2H_2O + 4e^-$

This is a *basic* solution, and no anions other than OH^- are present.

Statement 9

The applied potential necessary to effect a nonspontaneous process in an electrolytic cell will always be at least as great as the ϵ_{cell} values used for galvanic cells. In practice, some "**overvoltage**" is always required; that is, extra work must be done beyond the minimum ϵ_{cell} values calculated from a table of reduction potentials. We would expect that when several species are present, those which require the least total work to oxidize and reduce will be oxidized and reduced.

Predicting electrolysis products (that is, predicting which species are oxidized and reduced most easily) can often be done safely and simply by **comparing $\epsilon°$ values** (Table 18.1), recalling that **the more positive $\epsilon°$ is for a half-reaction**, the easier is **reduction of the oxidized form**, and the more difficult is **oxidation of the reduced form**.

We must keep in mind, however, that because of the overvoltage possibilities we cannot rely strictly on $\epsilon°$ comparisons in making predictions. The Cl^- ion is a common exception: it is oxidized more easily than H_2O despite the fact that the $\epsilon°$ value for $Cl_2 \rightarrow 2Cl^-$ is more positive than that for $O_2 \rightarrow H_2O$.

Question 33

Predict:

a. the cathode process

b. the anode process

c. the cell reaction when aqueous $CdBr_2$ undergoes electrolysis.

Answer 33

a. $Cd^{2+} + 2e^- \rightarrow Cd$

Possibilities	$e°$ value
$Cd^{2+} + 2e^- \rightarrow Cd$	-0.40v
$2H_2O + 2e^- \rightarrow H_2 + 2OH^-$	-0.83v

Thus $\epsilon°_{Cd^{2+}} > \epsilon°_{H_2O}$ and Cd^{2+} is easier to reduce.

b. $2Br^- \rightarrow Br_2 + 2e^-$

Possibilities	$e°$ value
$2Br^- \rightarrow Br_2 + 2e^-$	+1.06v
$2H_2O \rightarrow O_2 + 2H^+ + 2e^-$	+1.23v

Thus $\epsilon°_{Br^-} < \epsilon°_{H_2O}$ and Br^- is easier to *oxidize* than H_2O.

c. $Cd^{2+} + 2Br^- \rightarrow Cd + Br_2$

Simply add the two half-cell processes.

Question 34

Predict:

a. the cathode process
b. the anode process
c. the cell reaction
when aqueous KBr undergoes electrolysis.

Answer 34

a. $2H_2O + 2e^- \rightarrow H_2 + 2OH^-$

$\epsilon°_{H_2O} > \epsilon°_{K^+}$

b. $2Br^- \rightarrow Br_2 + 2e^-$

$\epsilon°_{Br^-} < \epsilon°_{H_2O}$

c. $2H_2O + 2Br^- \rightarrow Br_2 + H_2 + 2OH^-$

Question 35

Predict:

a. the cathode process

b. the anode process

c. the cell reaction

when aqueous NiF_2 undergoes electrolysis.

Answer 35

a. $Ni^{2+} + 2e^- \rightarrow Ni$

$\epsilon°_{Ni^{2+}} > \epsilon°_{H_2O}$

b. $2H_2O \rightarrow O_2 + 2H^+ + 2e^-$

$\epsilon°_{H_2O} < \epsilon°_{F^-}$

c. $Ni^{2+} + 2H_2O \rightarrow O_2 + 2H^+ + Ni$

Question 36

An aqueous solution of $MgSO_4$ undergoes electrolysis. Predict:

a. the cathode process

b. the anode process

c. the cell reaction

Answer 36

a. $2H_2O + 2e^- \rightarrow H_2 + 2OH^-$

b. $2H_2O \rightarrow O_2 + 4H^+ + 4e^-$

c. $2H_2O \rightarrow 2H_2 + O_2$

This is the electrolysis of water. Observe that when we added the two half-reactions, $4OH^- + 4H^+$ appeared as products. These would not coexist but would react to form more H_2O, so they do not appear in the final cell reaction.

Question 37

An aqueous solution of $MgCl_2$ undergoes electrolysis. Predict:

a. the cathode process

b. the anode process

c. the cell reaction

Answer 37

a. $2H_2O + 2e^- \rightarrow H_2 + 2OH^-$

b. $2Cl^- \rightarrow Cl_2 + 2e^-$

The Cl^- ion is an exception. Even though $\epsilon^\circ_{Cl^-} > \epsilon^\circ_{H_2O}$, Cl^- is reduced.

c. $2H_2O + 2Cl^- \rightarrow Cl_2 + H_2 + 2OH^-$

Question 38

A solution of $MgBr_2$ dissolved in 1 M HBr undergoes electrolysis. Predict:

a. the cathode process

b. the anode process

c. the cell reaction

Answer 38

a. $2H^+ + 2e^- \rightarrow H_2$

$\epsilon^\circ_{H^+} > \epsilon^\circ_{Mg^{2+}}$ (acid solution)

b. $2Br^- \rightarrow Br_2 + 2e^-$

c. $2H^+ + 2Br^- \rightarrow Br_2 + H_2$

Question 39

Predict the cell reaction when molten NaCl undergoes electrolysis.

Answer 39

$2Na^+ + 2Cl^- \rightarrow 2Na + Cl_2$

No H_2O, H^+, or OH^- to consider.

Statement 10

The **quantity of electrolysis products** can be predicted by using **Faraday's Law**:

$$\text{\# moles product} = \frac{q}{nF}$$

where q is the number of coulombs of electricity passed through the solution (or molten salt), n is the number of electrons exchanged in the balanced cell reaction, and F is a constant called "the faraday" (96,494 coulombs). The faraday is simply the number of coulombs equivalent to **one mole of electrons** (6.023×10^{23} electrons = 96.494 coulombs = 1 faraday). The number of coulombs may be determined by observing the length of time (in seconds), t, that a measured current flow (in amperes), i, takes place:

$$q = it$$

Question 40

In the electrolysis of aqueous $NiBr_2$ 8021 coulombs of electricity pass through the solution.

a. How many moles of Ni metal will be produced?

b. How many grams of Br_2 will be produced?

Answer 40

a. 0.04156 mole Ni

OEIOU: $\text{\# moles Ni} = \frac{q}{nF}$

$$= \frac{8.021 \text{ C}}{2(96494 \text{ C/mole})}$$

$$= 0.04156 \text{ mole}$$

b. 6.642 g Br_2

$$\text{\# moles Br}_2 = \frac{q}{nF}$$

$$\left(\frac{g}{GMW}\right)_{Br_2} = \frac{q}{nF}$$

OEIOU: $g_{Br_2} = \frac{q}{nF} \times (GMW)_{Br_2}$

$$= \frac{8.021 \text{ C}}{2(96494 \text{ C/mole})} \times 159.8 \text{ g/mole}$$

$$= 6.642 \text{ g}$$

PART 18—OXIDATION-REDUCTION PROCESSES

Question 41

How many grams of silver are deposited when 8.00 amperes of current is passed through a AgNO$_3$ solution for 6480 seconds?

Answer 41

58.0 g Ag

$$\text{\# moles Ag} = \frac{q}{nF}$$

$$(g/GAW)_{Ag} = \frac{it}{nF}$$

OEIOU:
$$g_{Ag} = \frac{it}{nF} \times (GAW)_{Ag}$$

$$= \frac{(8.00 \text{ amp})(6480 \text{ sec})}{1(96{,}494 \text{ amp sec/mole})} \times 107.87 \text{ g/mole}$$

$$= 58.0 \text{ g}$$

Note: 1 coulomb = 1 amp sec

Question 42

How long must a current of 5.00 amperes pass through a solution of AuCl$_3$ to plate out 10.0 g of Au metal?

Answer 42

980 sec

$$\text{OEIOU: } t = (g/GAW)_{Au} \times \frac{nF}{i}$$

Question 43

In the electrolysis of a solution of lead acetate, 9687 coulombs of electricity deposited 10.4 grams of Pb metal. Calculate the gram equivalent weight of Pb based on these data.

Answer 43

104 g/GEW

$$\text{\# moles Pb} = q/nF$$

PART 18—OXIDATION-REDUCTION PROCESSES

$$\left(\frac{g}{GAW}\right)_{Pb} = q/nF$$

$$\left(\frac{g}{GEW \times n_c}\right)_{Pb} = q/nF$$

$$\text{OEIOU:} \quad (GEW)_{Pb} = \frac{nF}{q} \times \left(\frac{g}{n_c}\right)_{Pb}$$

$$= \frac{2(96494 \text{ C/mole})}{9687 \text{ C}} \times \frac{10.4 \text{ g}}{2 \text{ GEW/mole}}$$

$$= 104 \text{ g/GEW}$$

Question 44

The Hall process for producing aluminum metal requires the electrolysis of molten Al_2O_3. How many gram equivalent weights of Al metal will be produced by passing 10.0 amperes of current through the molten salt for 2.00 hours?

Answer 44

0.746 GEW

$$\# \text{moles} = \frac{it}{nF}$$

$$\frac{\#(GEW)}{n_c} = \frac{it}{nF}$$

$$\text{OEIOU:} \quad \#(GEW) = \frac{it}{nF} \times n_c$$

$$= \frac{(10.0 \text{ amp})(2.00 \text{ hr} \times 3600 \text{ sec/hr})}{3(96494 \text{ C/mole})} \times 3 \text{ GEW/mole}$$

$$= 0.746 \text{ GEW}$$

TESTING YOUR MASTERY—Part 18
Oxidation-Reduction Processes

Question 45

In each of the following cases, predict which species would have the greater affinity for electrons under standard conditions.

a. Pb^{2+} or Fe^{3+}

b. Cl_2 or Ag^+

c. O_2 or Cd^{2+}

Answer 45

a. Fe^{3+}

b. Cl_2

c. O_2

Compare ϵ° values for each pair. The more positive ϵ° value corresponds to the greater affinity for electrons.

Question 46

A Pb strip is dipped into a 1.00 M $Pb(NO_3)_2$ solution in one side of a cell, and a silver strip is dipped into a 1.00 M $AgNO_3$ solution in the other side. If the Pb and Ag strips are connected by an external wire, show an equation for the reduction process that takes place.

Answer 46

$Ag^+ + e^- \rightleftarrows Ag$

More positive ϵ° value for the Ag^+ reduction.

Question 47

Which of these reductions will take place more easily?

a. $2e^- + 2H_2O \rightarrow H_2 + 2OH^-$

b. $2e^- + Ca^{2+} \rightarrow Ca$

Answer 47

Reduction (a), the reduction of water, takes place more easily.

More positive ϵ° value for the reduction of H_2O.

Question 48

Which of these reactions will proceed as shown under standard conditions?

a. $Sn^{2+} + Ni \rightleftarrows Ni^{2+} + Sn$

b. $2Al^{3+} + 3H_2 \rightleftarrows 6H^+ + 2Al$

c. $2MnO_4^- + 10Cl^- + 16H^+ \rightleftarrows 5Cl_2 + 2Mn^{2+} + 8H_2O$

Answer 48

(a) and (c) only.

> The reduced species on the left in (a) and (c) are in half-reactions which lie *above* the half-reactions containing the species they would be reducing in the activity series.

Question 49

Which of these reagents could oxidize Ni metal to Ni^{2+}?

$$I_2, \quad Cd^{2+}, \quad O_2$$

Answer 49

I_2 or O_2.

> Both I_2 and O_2 are oxidized forms appearing *below* Ni in the activity series.

Question 50

Which of these metals should dissolve in HCl?

$$Ag, \quad Cu, \quad Cd$$

Answer 50

Cd only.

> Cd is a reduced species *above* H^+ in the activity series.

Question 51

Arrange these reagents in order of decreasing oxidizing strength in acid solution.

$$I_2, \; O_2, \; Pb^{2+}, \; Fe^{3+}$$

Answer 51

$$O_2 > I_2 > Fe^{3+} > Pb^{2+}$$

Reverse order of appearance as oxidized forms in activity series. (Order of decreasing positive values for $\epsilon°$.)

Question 52

Arrange these reagents in order of decreasing strength as reducing agents in water solution.

$$Cl^-, \; H_2, \; K, \; I^-$$

Answer 52

$$K > H_2 > I^- > Cl^-$$

Same order of appearance as reduced forms in activity series. (Order of increasing $\epsilon°$ values.)

Question 53

On which of the following metals should Pb^{2+} plate out as Pb metal when a strip of the metal in question is dipped into $Pb(NO_3)_2$ solution?

$$Ag, \; Cu, \; Ni, \; Cd$$

Answer 53

Ni and Cd.

Pb^{2+} lies below Ni and Cd in the activity series, but above Ag and Cu.

Question 54

Complete and balance the following equations:

a. $Cl_2 + Fe^{2+} \rightleftarrows Fe^{3+} + Cl^-$

b. $I_2 + Cr^{3+} \rightleftarrows Cr_2O_7^{2-} + I^-$ (in acid solution)

PART 18—OXIDATION-REDUCTION PROCESSES

Answer 54

a. $Cl_2 + 2Fe^{2+} \rightleftarrows 2Fe^{3+} + 2Cl^-$

b. $3I_2 + 2Cr^{3+} + 7H_2O \rightleftarrows Cr_2O_7^{2-} + 6I^- + 14H^+$

Question 55

Complete and balance the following equations:

a. $Cd + MnO_4^- \rightleftarrows Mn^{2+} + Cd^{2+}$ (acid solution)

b. $Cd + MnO_4^- \rightleftarrows MnO_2 + Cd^{2+}$ (basic solution)

Answer 55

a. $5Cd + 2MnO_4^- + 16H^+ \rightleftarrows 2Mn^{2+} + 5Cd^{2+} + 8H_2O$

b. $3Cd + 2MnO_4^- + 4H_2O \rightleftarrows 2MnO_2 + 2Cd^{2+} + 8OH^-$

Question 56

Complete and balance these equations:

a. $Cu + NO_3^- \rightleftarrows Cu^{2+} + NO$ (acid solution)

b. $ClO_3^- + NO \rightleftarrows Cl_2 + NO_3^-$ (acid solution)

Answer 56

a. $5Cu + 2NO_3^- + 12H^+ \rightleftarrows 5Cu^{2+} + 2NO + 6H_2O$

b. $6ClO_3^- + 10NO + 2H_2O \rightleftarrows 3Cl_2 + 10NO_3^- + 4H^+$

Question 57

Complete and balance these equations:

a. $H_2O_2 + Mn^{2+} \rightleftarrows MnO_2 + H_2O$ (acid solution)

b. $O_2 + NO_2^- \rightleftarrows NO_3^- + OH^-$ (basic solution)

Answer 57

a. $H_2O_2 + Mn^{2+} \rightleftarrows MnO_2 + 2H^+$

b. $O_2 + 2NO_2^- \rightleftarrows 2NO_3^-$

Question 58

Write the cell diagram for each of these cell reactions:

a. $Mg + Ni^{2+} \rightleftarrows Ni + Mg^{2+}$

b. $Zn + 2H^+ \rightleftarrows H_2 + Zn^{2+}$

Answer 58

a. $Mg \mid Mg^{2+} \parallel Ni^{2+} \mid Ni$

b. $Zn \mid Zn^{2+} \parallel H^+ \mid H_2 Pt$

Question 59

Write the cell reactions corresponding to these cell diagrams:

a. $Pb \mid Pb^{2+} \parallel Ag^+ \mid Ag$

b. $Mg \mid Mg^{2+} \parallel Sn^{2+} \mid Sn$

Answer 59

a. $Pb + 2Ag^+ \rightleftarrows 2Ag + Pb^{2+}$

b. $Mg + Sn^{2+} \rightleftarrows Sn + Mg^{2+}$

Question 60

Calculate ϵ°_{cell} for these oxidation-reduction processes:

a. $O_2 + 2Cu + 4H^+ \rightleftarrows Cu^{2+} + 2H_2O$

b. $Cl_2 + 2I^- \rightleftarrows I_2 + 2Cl^-$

c. $2Al + 3PbSO_4 \rightleftarrows Pb + SO_4^{2-} + 2Al^{3+}$

Answer 60

a. +0.89v

b. +0.82v

c. +1.30v

OEIOU: $\epsilon^\circ_{cell} = \epsilon^\circ_{cat} - \epsilon^\circ_{An}$

Decide which is cathode by observing which species is reduced in the equation.

Question 61

Which of these reactions are spontaneous as written, based on ϵ°_{cell} values?

a. $H_2 + Sn^{2+} \rightleftarrows Sn + 2H^+$

b. $4I^- + O_2 + 4H^+ \rightleftarrows 2I_2 + 2H_2O$

c. $Cr_2O_7^{2-} + 6Fe^{2+} + 14H^+ \rightleftarrows 2Cr^{3+} + 6Fe^{3+} + 7H_2O$

Answer 61

(b) and (c) are spontaneous, since ϵ°_{cell} is *positive* for them.

OEIOU: $\epsilon^\circ_{cell} = \epsilon^\circ_{cat} - \epsilon^\circ_{An}$

(Do calculations based on cell *as written*.)

Question 62

Calculate ϵ_{cell} for the reaction $Sn + Pb^{2+} \rightleftarrows Pb + Sn^{2+}$

a. when $[Pb^{2+}]$ is 1.00 M and $[Sn^{2+}]$ is 1.00 M.

b. when $[Pb^{2+}]$ is 0.100 M and $[Sn^{2+}]$ is 0.500 M.

Answer 62

a. +0.01 v

OEIOU: $\epsilon_{cell} = \epsilon^\circ_{cell} = \epsilon^\circ_{Pb} - \epsilon^\circ_{Sn}$

b. −0.01 v

OEIOU: $\epsilon_{cell} = \epsilon^\circ_{cell} - \dfrac{0.0591}{n} \log \dfrac{[Sn^{2+}]}{[Pb^{2+}]}$

Question 63

Compare ϵ_{cell} values for the reaction

$5Cd + 2MnO_4^- + 16H^+ \rightleftarrows 2Mn^{2+} + 5Cd^{2+} + 8H_2O$

when:

a. all ionic species are 1.00 M.

b. the H^+ concentration is 0.10 M, but all other ionic species are 1.00 M.

Answer 63

a. +1.92v

OEIOU: $\epsilon_{cell} = \epsilon°_{cell} = \epsilon°_{MnO_4^-} - \epsilon°_{Cd}$

b. +0.97v

OEIOU: $\epsilon_{cell} = \epsilon°_{cell} - \dfrac{0.0591}{n} \log \dfrac{[Mn^{2+}]^2 [Cd^{2+}]^5}{[MnO_4^-]^2 [H^+]^{16}}$

Question 64

Calculate ϵ_{cell} for the following cell at 25°C:

$$Ni \mid Ni^{2+}\ (0.0684\ M)\ \|\ Ag^+(0.0200\ M) \mid Ag$$

Answer 64

+0.98v

OEIOU: $\epsilon_{cell} = \epsilon°_{cell} - \dfrac{0.0591}{n} \log \dfrac{[Ni^{2+}]}{[Ag^+]^2}$

Question 65

ϵ_{cell} for the reaction

$$Ni + 2Ag^+ \rightleftarrows 2Ag + Ni^{2+}$$

is +1.05v.

a. What is K_e at 25°C?

b. Does the reaction go essentially to completion?

Answer 65

a. 3.4×10^{35}

OEIOU: $K_e = \text{antilog} \left[\dfrac{n\ \epsilon°_{cell}}{0.0591} \right]$

b. Yes. K_e is very large.

Question 66

What is the molar ratio of Cd^{2+} to Ni^{2+} at equilibrium when Cd metal is dissolved in $Ni(NO_3)_2$ solution?

Answer 66

1.2×10^5

OEIOU: $\dfrac{[Cd^{2+}]}{[Ni^{2+}]} = \text{antilog}\left[\dfrac{-n\, \epsilon^\circ_{cell}}{0.0591}\right]$

Question 67

What are the products of electrolysis for molten RbI?

Answer 67

Rb metal and I_2

Cell reaction: $2Rb^+ + 2I^- \rightarrow 2Rb + I_2$

Question 68

What are the products of electrolysis for aqueous RbI?

Answer 68

H_2 (and OH^-) and I_2

Cell reaction: $2H_2O + 2I^- \rightarrow H_2 + 2OH^- + I_2$

Question 69

What are the products of electrolysis for aqueous KOH?

Answer 69

H_2 gas and O_2 gas

Cell reaction: $2H_2O \rightarrow 2H_2 + O_2$

Question 70

Write an equation for:

a. the cathode process

b. the anode process

c. the cell reaction

when aqueous $SnCl_2$ undergoes electrolysis in acid solution.

Answer 70

a. cathode: $Sn^{2+} + 2e^- \rightarrow Sn$

b. anode: $2Cl^- \rightarrow Cl_2 + 2e^-$

c. cell: $Sn^{2+} + 2Cl^- \rightarrow Sn + Cl_2$

Question 71

Write an equation for:

a. the cathode process

b. the anode process

c. the cell reaction

when aqueous HBr undergoes electrolysis.

Answer 71

a. cathode: $2H^+ + 2e^- \rightarrow H_2$

b. anode: $2Br^- \rightarrow Br_2 + 2e^-$

c. cell: $2H^+ + 2Br^- \rightarrow H_2 + Br_2$

Question 72

In the electrolysis of molten $CaCl_2$, 1.00×10^4 coulombs of electricity pass through the solution.

a. How many grams of Ca metal are produced?

b. How many moles of Cl_2 gas are formed?

Answer 72

a. 2.08 g Ca

OEIOU: $g_{Ca} = \dfrac{q}{nF} \times (GAW)_{Ca}$

b. 0.0518 mole Cl_2

OEIOU: # moles $Cl_2 = \dfrac{q}{nF}$

Question 73

How many grams of Pb metal will be deposited when 9.75 amperes of current is passed through a lead nitrate solution for 7500 seconds?

Answer 73

78.5 g Pb

OEIOU: $g_{Pb} = \dfrac{it}{nF} \times (GAW)_{Pb}$

Question 74

Sodium metal is produced commercially by the Downs process—the electrolysis of molten NaCl. How many seconds are required to form one mole (22.99 g) of Na metal if the current is 10.0 amperes?

Answer 74

9.65×10^3 sec

OEIOU: $t = \text{\# moles Na} \times \dfrac{nF}{i}$

PART 19
COMPLEX IONS

Statement 1

A great variety of **complex ions** is formed by the coordination of several **ligands** (:L)—molecules or anions having one or more pairs of unshared electrons; with a **metal ion** (M^{m+})—a cation derived from a metal atom, most commonly, an ion of a transition metal.

The number of points of coordination between ligands and metal in a complex is known as the **coordination number**, C.N., of the metal in the complex. For simple complexes of the form ML_x^{n+}, in which each ligand is attached to M at only one point, C.N. = x.

$$\begin{array}{c} L \\ \downarrow \\ L \rightarrow M \leftarrow L \\ \uparrow \\ L \end{array}$$

Note: Not all complexes are charged (i.e., not all are ions). Nor must M necessarily be ionic; it can be an atom. In solution chemistry, however, most complex species encountered will be ionic.

Question 1

For the complex $Cu(NH_3)_4^{2+}$,

a. what is the ligand?

b. what is the C.N. of the central ion?

Answer 1

a. NH_3

b. four. Each $:NH_3$ molecule has one point of attachment to Cu^{2+}

Question 2

For the complex $Co(NH_3)_4Cl_2^+$,

a. identify the ligands.

b. give the C.N. of the central metal ion.

> **Answer 2**
>
> a. NH_3 and Cl^-
>
> b. six—four NH_3 molecules and two Cl^{-3} ions.

Statement 2

In some cases a ligand may be fairly large and have more than one point in the molecule where there are unshared electrons. We may represent a "bidentate" (a ligand having two possible points of attachment) by:

$$: \overset{\frown}{L \quad L} :$$

where the curved line represents a connecting string of atoms in the molecule. The resulting complex now has the form:

$$\begin{array}{c} L \\ \downarrow \\ L \rightarrow M \leftarrow L \\ \uparrow \\ L \end{array}$$

and is called a **chelate**.

Question 3

Which of these molecules might serve as a bidentate?

$CH_3CH_2CH_2-\ddot{N}H_2$

$\begin{array}{c} CH_3 \\ \diagdown \\ B-CH_3 \\ \diagup \\ CH_3 \end{array}$ $\ddot{N}H_2-CH_2CH_2-\ddot{N}H_2$

> **Answer 3**
>
> $NH_2-CH_2CH_2-NH_2$ ethylenediamine
>
> Only this molecule has *two* pairs of unshared electrons, one on each N atom.

Question 4

For the complex $Co(en)_2Cl_2^+$,

a. identify the ligands.
b. give the C.N. of the central metal ion.

(**Note:** en = ethylenediamine, $NH_2-CH_2CH_2-NH_2$.)

Answer 4

a. $NH_2-CH_2CH_2-NH_2$ and Cl^-

b. six. Each en molecule has *two* points of attachment, and each Cl^- has *one*:

$$(2 \times 2) + (2 \times 1) = 6$$

Question 5

Which of these complexes is a chelate?

$Co(NH_3)_6^{3+}$

$Co(en)_3^{3+}$

$Cr(H_2O)_4Cl_2^+$

Answer 5

$Co(en)_3^{3+}$

The other complexes have only "monodentate" ligands capable of only *one* point of attachment. The en ligand has *two* points of attachment, so forms a chelate.

Statement 3

A complex ion may be either the cation or anion of an ionic compound. When **writing formulas** for compounds involving complexes, the **primary coordination sphere**—ligands plus metal—is enclosed in **brackets**:

$[ML_x]Y$ complex as a cation

$M'[ML_x]$ complex as an anion

where Y is an anion and M' a cation.

Question 6

In the complex $[Co(NH_3)_4Cl(NO_2)]Br$, which species are:

a. inside the coordination sphere?

b. outside the coordination sphere?

> **Answer 6**
>
> a. inside: Co^{3+}, NH_3, Cl^-, NO_2^-
>
> b. outside: Br^-

Question 7

Write the formula for the complex having a barium ion outside the coordination sphere, and one Zn^{2+} plus four OH^- ions inside the sphere.

> **Answer 7**
>
> $Ba[Zn(OH)_4]$

Question 8

Write the formula for the compound having Cl^- ion(s) outside the coordination sphere, and one Co^{3+}, two NH_3, and two Cl^- inside the sphere.

> **Answer 8**
>
> $[Co(NH_3)_2Cl_2]Cl$

Statement 4

To name complex ions and their compounds, we follow systematic **rules of nomenclature**:

(1) Name the **cation** of the compound first, then the **anion**.

(2) When naming the complex, name the **ligands** first, then the **metal**.

(3) When naming the ligands in a complex, name **negative** ligands, then **neutral** ligands. Within each of these categories, name ligands in **alphabetic order**. (See Table 19.1 for ligand names.)

(4) Use the **prefixes** di-, tri-, tetra-, and so forth, before the names of simple (monodentate) ligands when more than one of a given kind is present. Use the prefixes bis-, tris-, tetrakis-, and so forth, for bidentates or higher ligands.

(5) To indicate the charge on the central metal, use a Roman numeral in parentheses after its name.

(6) If the complex is an anion, the ending -ate follows the name of the metal. Latin stems are often used for the metal in such cases.

TABLE 19.1 LIGAND NAMES

ammine	NH_3
aquo	H_2O
carbonyl	CO
ethylenediamine ("en")	$NH_2CH_2CH_2NH_2$
nitrosyl	NO
acetato	OAc^- $(CH_3-\overset{O}{\underset{\|}{C}}-O)$
bromo	Br^-
carbonato	CO_3^{2-}
chloro	Cl^-
cyano	CN^-
hydroxo	OH^-
iodo	I^-
nitrito	NO_2^-

Question 9

Name these complexes:

a. $Ag(NH_3)_2^+$

b. $Cu(NH_3)_4^{2+}$

Answer 9

a. diamminesilver(I) ion

b. tetramminecopper(II) ion

Note that there is no space between ligand name and metal name.

Question 10

Name these complexes:

a. $Cr(CN)_6^{3-}$

b. $Fe(CN)_5CO^{3-}$

Answer 10

a. hexacyanochromate(III) ion

b. pentacyanocarbonylironate(II) ion

or

pentacyanocarbonylferrate(II) ion

Question 11

Name these compounds:

a. $[Cr(H_2O)_4Cl_2]Br$

b. $[Cu(en)_2]SO_4$

c. $Na_2[ZnCl_4]$

Answer 11

a. dichlorotetraquochromium(III) bromide

b. bis(ethylenediamine)copper(II) sulfate

c. sodium tetrachlorozincate(II)

Question 12

Write the formula for:

a. hexamminecobalt(III) chloride

b. potassium pentacyanonitrosylcobaltate(III)

Answer 12

a. $[Co(NH_3)_6]Cl_3$

b. $K_2[Co(CN)_5NO]$

Statement 5

In an aqueous solution, a complex undergoes **stepwise dissociation**. For example:

$$ML_2 \rightleftharpoons ML + L \qquad K_1 = \frac{[ML][L]}{[ML_2]}$$

$$ML \rightleftharpoons M + L \qquad K_2 = \frac{[M][L]}{[ML]}$$

Knowing the equilibrium constant for each step, we know the extent to which each step of dissociation takes place. For the overall dissociation

$$ML_2 \rightleftharpoons M + 2L$$

we may also write:

$$K_{inst} = \frac{[M][L]^2}{[ML_2]}$$

where the equilibrium constant is called the **instability constant**. Some K_{inst} constants are given in Table 19.2.

TABLE 19.2 INSTABILITY CONSTANT VALUES

$Ag(NH_3)_2^+$	6.3×10^{-8}
$Ag(CN)_2^-$	1.0×10^{-21}
$Cu(NH_3)_4^{2+}$	4.6×10^{-14}
$Ni(NH_3)_4^{2+}$	5.0×10^{-8}
$Ni(CN)_4^{2-}$	1.0×10^{-22}

Question 13

For $Cu(NH_3)_4^{2+}$:

a. write the stepwise dissociation equilibria.
b. write the individual equilibrium expressions.
c. write the overall instability expression.

Answer 13

a. $Cu(NH_3)_4^{2+} \rightleftharpoons Cu(NH_3)_3^{2+} + NH_3$

$Cu(NH_3)_3^{2+} \rightleftharpoons Cu(NH_3)_2^{2+} + NH_3$

$$Cu(NH_3)_2^{2+} \rightleftharpoons Cu(NH_3)^{2+} + NH_3$$

$$Cu(NH_3)^{2+} \rightleftharpoons Cu^{2+} + NH_3$$

b. $K_1 = \dfrac{[Cu(NH_3)_3^{2+}][NH_3]}{[Cu(NH_3)_4^{2+}]}$

$K_2 = \dfrac{[Cu(NH_3)_2^{2+}][NH_3]}{[Cu(NH_3)_3^{2+}]}$

$K_3 = \dfrac{[Cu(NH_3)^{2+}][NH_3]}{[Cu(NH_3)_2^{2+}]}$

$K_4 = \dfrac{[Cu^{2+}][NH_3]}{[Cu(NH_3)^{2+}]}$

c. $K_{inst} = \dfrac{[Cu^{2+}][NH_3]^4}{[Cu(NH_3)_4^{2+}]}$

Question 14

Which of the following complexes is more stable:

$$Ag(NH_3)_2^+ \quad \text{or} \quad Ag(CN)_2^- \;?$$

Answer 14

$Ag(CN)_2^-$

K_{inst} is *smaller* for the more *stable* species as long as species having the same C.N. are being compared.

Question 15

Which of these complexes is less stable:

$$Cu(NH_3)_4^{2+} \quad \text{or} \quad Ni(NH_3)_4^{2+} \;?$$

Answer 15

$Ni(NH_3)_4^{2+}$

$K_{inst,\ Ni(NH_3)_4^{2+}} > K_{inst,\ Cu(NH_3)_4^{2+}}$

Statement 6

When calculating concentrations of individual species in a solution of a complex, conditions often exist that do not permit the same sort of simplifying assumptions that can be made for polyacids and polybases, because K_e values for several stepwise dissociations may have comparable values.

When a **large excess of ligand** is present, however, we are usually justified in assuming that essentially all of the metal is tied up in the most highly complexed form. Thus, the approximate concentration of the most highly complexed form is given by the following equation:

$$[ML_x] \approx C_M$$

where C_M is the formal concentration of dissolved metal ion, both complexed and uncomplexed. We may further assume in such cases of excess ligand:

$$[L] \approx C_L$$

where $[L]$ is the actual concentration of free ligand and C_L its formal concentration.

Question 16

What are the approximate concentrations of $Ag(NH_3)_2^+$ and NH_3 in a solution formed by mixing 0.0050 mole $AgNO_3$ and 1.0 mole NH_3 in 500 ml of water?

Answer 16

$[Ag(NH_3)_2^+] \approx 0.010 \text{ M} \qquad [NH_3] \approx 2.0 \text{ M}$

$[Ag(NH_3)_2^+] \approx C_M$

OEIOU: $[Ag(NH_3)_2^+] \approx \dfrac{\text{\# moles Ag}}{\text{\# L solution}}$

$[NH_3] \approx C_L$

OEIOU: $[NH_3] \approx \dfrac{\text{\# moles } NH_3}{\text{\# L solution}}$

Question 17

Calculate the approximate concentration of free silver ion in a solution obtained by mixing 0.010 mole of $AgNO_3$ and 2.0 moles NH_3 in a liter of water.

Answer 17

1.6×10^{-10} M

$$Ag(NH_3)_2^+ \rightleftharpoons Ag^+ + 2NH_3$$

$$K_{inst} = \frac{[Ag^+][NH_3]^2}{[Ag(NH_3)_2^+]} \approx \frac{[Ag^+] C_L^2}{C_M}$$

$$[Ag^+] \approx K_{inst} \frac{C_M}{C_L^2}$$

OEIOU: $[Ag^+] \approx K_{inst} \dfrac{(\# \text{moles Ag}/\# \text{L solution})}{(\# \text{moles } NH_3/\# \text{L solution})^2}$

Question 18

How much uncomplexed Ni^{2+} ion remains in solution when 0.050 mole of $NiSO_4$ and 1.50 moles of NaCN are dissolved in a liter of water?

Answer 18

9.9×10^{-25} M

$$Ni(CN)_4^{2-} \rightleftharpoons Ni^{2+} + 4CN^-$$

$$[Ni^{2+}] \approx K_{inst} \frac{C_M}{C_L^4}$$

OEIOU: $[Ni^{2+}] \approx K_{inst} \dfrac{(\# \text{moles Ni}/\# \text{L solution})}{(\# \text{moles CN}/\# \text{L solution})^4}$

Statement 7

The formation of a complex may radically **increase the solubility** of a sparingly soluble salt having an ion in common with the complex.

For example, if a potential ligand, L, is added to a solution in equilibrium with a precipitated salt, MX, capable of forming a complex, ML_2^+, **competing equilibria** exist:

$$MX \rightleftharpoons M^+ + X^- \quad \text{and} \quad ML_2^+ \rightleftharpoons M^+ + 2L$$

The extent to which additional MX dissolves depends on the result of the competition of X^- and L for the M^+ ion.

PART 19—COMPLEX IONS

If most of the dissolved M^+ is tied up as ML_2^+:

$$s_{MX} \approx [ML_2^+]$$

where s_{MX} is the molar solubility of ML. And for every MX dissolving we get one X^-, so:

$$s_{MX} = [X^-]$$

Thus:

$$K_{sp} = [M^+]s_{MX} \quad \text{and} \quad K_{inst} \approx \frac{[M^+]\,C_L^2}{s_{MX}}$$

Question 19

A solution in equilibrium with AgCl is made 0.80 M in NH_3. Calculate the molar solubility of AgCl in this solution.

$(K_{sp,AgCl} = 1.8 \times 10^{-10})$

Answer 19

4.3×10^{-2} M

$$K_{sp} = [Ag^+]s_{AgCl} \quad \text{and} \quad K_{inst} \approx \frac{[Ag^+]\,C_L^2}{s_{AgCl}}$$

Thus: $\dfrac{K_{sp}}{s_{AgCl}} \approx \dfrac{K_{inst}\, s_{AgCl}}{C_L^2}$

OEIOU: $s_{AgCl} \approx \sqrt{\dfrac{K_{sp}\,C_L^2}{K_{inst}}}$

$$= \sqrt{\frac{(1.8 \times 10^{-10})(8.0 \times 10^{-1})^2}{6.3 \times 10^{-8}}}$$

$$= 4.3 \times 10^{-2} \text{ M}$$

Question 20

Compare the solubility of NiS in pure water with its solubility in 1.0 M NaCN solution.

$(K_{sp,NiS} = 1.8 \times 10^{-21})$

Answer 20

In NaCN solution: 4.2 M

In pure water: 4.2×10^{-11} M

In NaCN solution:

$$\frac{K_{sp}}{s_{NiS}} \approx \frac{K_{inst}\, s_{NiS}}{C_L^4}$$

OEIOU: $s_{NiS} \approx \sqrt{\dfrac{K_{sp} C_L^4}{K_{inst}}}$

$$= \sqrt{\frac{(1.8 \times 10^{-21})(1.0)^4}{1.0 \times 10^{-22}}}$$

$$= 4.2 \text{ M}$$

In pure water:

$$K_{sp} = [Ni^{2+}]S^{2-}$$

$$= s_{NiS}^2$$

OEIOU: $s_{NiS} = \sqrt{K_{sp}}$

$$= \sqrt{1.8 \times 10^{-21}}$$

$$= 4.2 \times 10^{-11} \text{ M}$$

TESTING YOUR MASTERY—Part 19   Complex Ions

Question 21

For the complex compound $[Al(OH)(H_2O)_5]SO_4$:

a. What is the name of the compound?
b. Identify the ligands.
c. What is the coordination number of the central ion?

Answer 21

a. hydroxopentaquoaluminum(III) sulfate
b. OH^- and H_2O
c. 6

Question 22

Name these coordination compounds and show the C.N. of the central metal ion.

a. $[Pt(NH_3)_4(H_2O)Cl] Br_3$
b. $(NH_4)_2 [CeCl_2(NO_3)_4]$
c. $[Pt(en)(NH_3)_2] Cl_2$

Answer 22

a. chlorotetrammineaquoplatinum(IV) bromide
   C.N. = 6

b. ammonium dichlorotetranitratocerate(IV)
   C.N. = 6

c. diammine(ethylenediamine)platinum(II) chloride
   C.N. = 4

Question 23

In Question 22, what species are inside the coordination sphere?

Answer 23

a. Pt^{4+} NH_3 H_2O Cl^-
b. Ce^{4+} Cl^- NO_3^-
c. Pt^{2+} en NH_3

Question 24

Are any of the complexes in Question 22 chelates?

Answer 24

Yes. The cation in (c) is a chelate; it contains the bidentate "en" ligand.

Question 25

Show the formulas corresponding to these names.

a. tetracarbonylnickel(0)

b. potassium hexacyanoferrate(III)

c. tetrammineplatinum(II) tetrachloroplatinate(II)

Answer 25

a. $Ni(CO)_4$

b. $K_3[Fe(CN)_6]$

c. $[Pt(NH_3)_4][PtCl_4]$

Question 26

For $NiCl_4^{2-}$,

a. write the stepwise dissociation equilibria.

b. write the overall instability expression.

Answer 26

a. $NiCl_4^{2-} \rightleftarrows NiCl_3^- + Cl^-$

$NiCl_3^- \rightleftarrows NiCl_2 + Cl^-$

$NiCl_2 \rightleftarrows NiCl^+ + Cl^-$

$NiCl^+ \rightleftarrows Ni^{2+} + Cl^-$

b. $K_{inst} = \dfrac{[Ni^{2+}][Cl^-]^4}{[NiCl_4^{2-}]}$

Question 27

Arrange these complexes in order of decreasing stability:

$Cu(NH_3)_4^{2+}$

$Ni(NH_3)_4^{2+}$

$Ni(CN)_4^{2-}$

PART 19—COMPLEX IONS

Answer 27

$$Ni(CN)_4^{2-} > Cu(NH_3)_4^{2+} > Ni(NH_3)_4^{2+}$$

Arrange in order of increasing K_{inst} values.

Question 28

Find the approximate concentrations of $Cu(NH_3)_4^{2+}$ and NH_3 in a solution formed by mixing 0.0020 mole $CuSO_4$ and 0.50 mole NH_3 in 500.0 ml of water.

Answer 28

$$[Cu(NH_3)_4^{2+}] \approx 0.0010 \text{ M}$$

OEIOU: $[Cu(NH_3)_4^{2+}] \approx \dfrac{\text{\# moles Cu}^{2+}}{\text{\# L solution}}$

$$[NH_3] \approx 0.25 \text{ M}$$

OEIOU: $[NH_3] \approx \dfrac{\text{\# moles NH}_3}{\text{\# L solution}}$

Question 29

What is the approximate concentration of free Cu^{2+} in the solution in Question 28?

Answer 29

1.2×10^{-14} M

OEIOU: $[Cu^{2+}] \approx \dfrac{(\text{\# moles Cu}/\text{\# L solution})}{(\text{\# moles NH}_3/\text{\# L solution})^4}$

Question 30

What is the molar solubility of AgI in 0.050 M NaCN solution?

$(K_{sp}$ for $AgI = 8.5 \times 10^{-17})$

Answer 30

14 mole/L

OEIOU: $s_{AgBr} \approx \sqrt{\dfrac{K_{sp} C_L^2}{K_{inst,\ Ag(CN)_2^-}}}$

Question 31

What is the solubility of Ag_2CrO_4 in a 1.00 M NH_3 solution?

(K_{sp} for $Ag_2CrO_4 = 1.9 \times 10^{-12}$)

Answer 31

3.1×10^{-2} M

OEIOU: $s \approx \sqrt[3]{\dfrac{K_{sp} C_L^4}{K_{inst,\ Ag(NH_3)_2}}}$

PART 20
ORGANIC CHEMICALS

Statement 1

Organic chemistry is the chemistry of carbon-containing compounds. To call CH_3OH "organic" while calling Na_2CO_3 "inorganic" is to be arbitrary. There is, nonetheless, a generally agreed upon body of knowledge encompassed by organic chemistry, and our mastery of it should not be hampered by semantics.

The **alkanes** are the simplest family of organic compounds, in terms both of structure and of chemistry. Alkanes are **saturated hydrocarbons**, compounds composed entirely of carbon and hydrogen and containing only **single sigma bonds**. Relationships among the hydrocarbons may be shown schematically:

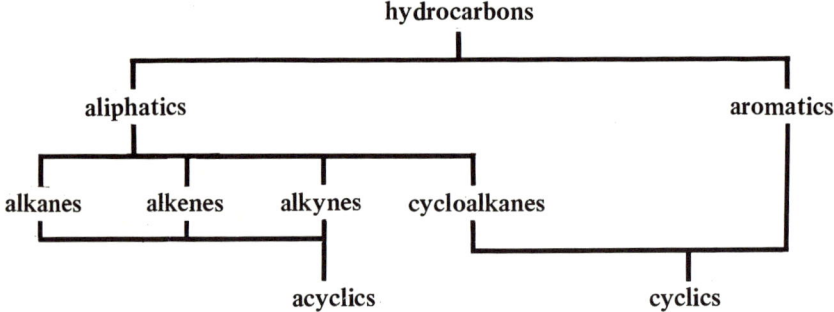

An infinite number of different alkanes may exist. These include both open chain (or **acyclic**) and **cyclic** types. The molecular formula of an alkane tells us which variety it may be:

$$\text{open chain alkanes} \qquad C_nH_{2n+2}$$

$$\text{cycloalkanes} \qquad C_nH_{2n}$$

Examples:

pentane
C_5H_{12}
(open chain)

cyclopentane
C_5H_{10}
(cyclic)

The name of an alkane is determined by the number of carbons it contains. If the carbons loop back to form a cyclic structure, the prefix cyclo- is placed before the name. See Table 20.1 for the names of alkanes having up to twelve carbons.

The chemist seldom depicts the "expanded" type of structures used in the preceding examples but more commonly **condenses** the formulas by denoting directly after each C symbol the number of hydrogens connected to it. Further condensation may be accomplished by using subscripts to show the number of repeating methylene ($-CH_2-$) units in a chain. It is important to be able to recognize a compound regardless of the convention chosen for depicting it. Note especially that the identity of an alkane is determined by the number of atoms present of each kind and their order of connection to one another regardless of how the picture is drawn.

Example

All of these are valid representations of **pentane**.

$$H-\underset{\underset{H}{|}}{\overset{\overset{H}{|}}{C}}-\underset{\underset{H}{|}}{\overset{\overset{H}{|}}{C}}-\underset{\underset{H}{|}}{\overset{\overset{H}{|}}{C}}-\underset{\underset{H}{|}}{\overset{\overset{H}{|}}{C}}-\underset{\underset{H}{|}}{\overset{\overset{H}{|}}{C}}-H \qquad CH_3-CH_2-CH_2\,CH_2-CH_3 \qquad CH_3(CH_2)_3CH_3$$

$$CH_3 \diagdown_{CH_2} \diagup^{CH_2} \diagdown_{CH_2} \diagup^{CH_3} \qquad \underset{CH_2CH_2CH_3}{\overset{CH_3CH_2}{|}} \qquad \underset{CH_2CH_2CH_2}{\overset{CH_3 \quad CH_3}{|\quad\quad|}} \qquad CH_3CH_2CH_2CH_2CH_3$$

Question 1

Which of these hydrocarbons may be alkanes?

a. $C_{20}H_{42}$

b. $C_{20}H_{38}$

c. $C_{18}H_{38}$

> **Answer 1**
>
> a and c only.
>
> > These have the formula C_nH_{2n+2}. Compound (b) has the formula C_nH_{2n-2}, so can be neither an open chain nor a cycloalkane.

Question 2

Which of these alkanes are cycloalkanes?

a. $C_{13}H_{28}$

b. $C_{10}H_{20}$

c. $C_{14}H_{28}$

Answer 2

b and c only.

These have the formula C_nH_{2n}, so are cycloalkanes. Compound (a) has the formula C_nH_{2n+2}, so must be acyclic.

Question 3

What type of hybridization is involved for the C atoms in the alkanes? (Refer back to Part 4, Statement 8, if necessary, for a review of valence-bond theory.)

Answer 3

sp^3 hybridization

Carbon has 4 single bonds to hydrogens and/or other carbons, requiring 4 hybrid orbitals. Only sp^3 hybridization would allow this.

Question 4

Name these alkanes.

a. C_9H_{20}

b. C_9H_{18}

c. C_6H_{14}

d. C_7H_{14}

Answer 4

a. nonane

b. cyclononane

c. hexane

d. cycloheptane

Consult Table 20.1. Use the cyclo- prefix for alkanes with C_nH_{2n} formulas.

Question 5

Identify (name) these alkanes.

a. $CH_3(CH_2)_6CH_3$

b. $CH_2\begin{smallmatrix}CH_2-CH_2\\CH_2-CH_2\end{smallmatrix}CH_2$

c. $CH_3CH_2\underset{\underset{CH_2CH_2CH_2CH_3}{|}}{}$

d. $CH_3\diagdown_{CH_2}\diagup^{CH_2}\diagdown_{CH_2}\diagup^{CH_2}\diagdown CH_3$

Answer 5

a. hexane

b. cyclohexane

c. hexane

d. hexane

Remember: Hexane is hexane no matter how we choose to draw it.

Statement 2

Alkenes are aliphatic hydrocarbons which contain one or more **double bonds**, each double bond consisting of one **sigma** and one **pi** bond. They are said to be **unsaturated** hydrocarbons inasmuch as they contain less than the maximum number of hydrogens that could be accommodated by the carbons present. The general formulas for the alkenes and cycloalkenes are:

open chain alkenes $\qquad C_nH_{2n}$

cycloalkenes $\qquad C_nH_{2n-2}$

Nomenclature for the simple alkenes is also based on the names given in Table 20.1, but the suffix **-ene** rather than -ane is used. Alkene nomenclature also requires that when there may be confusion otherwise, a **numerical prefix** be used to show the **location of the double bond** along the carbon chain. The carbons are numbered from the end of the chain that gives the **lowest number** to a carbon that is doubly bonded, and the number of only one of the pair of carbons involved in the bond is needed.

Examples

$CH_2=CHCH_2CH_2CH_3$ $\qquad$ $CH_2CH=CHCH_2CH_3$ $\qquad$ cyclopentene

1-pentene $\qquad\qquad\qquad$ 2-pentene

PART 20–ORGANIC CHEMICALS

TABLE 20.1 ALKANE NAMES

Molecular Formula*	Name
CH_4	methane
C_2H_6	ethane
C_3H_8	propane
C_4H_{10}	butane
C_5H_{12}	pentane
C_6H_{14}	hexane
C_7H_{16}	heptane
C_8H_{18}	octane
C_9H_{20}	nonane
$C_{10}H_{22}$	decane
$C_{11}H_{24}$	undecane
$C_{12}H_{26}$	dodecane

*Cycloalkanes have 2 fewer H atoms per molecule.

Question 8

Which of these may be alkenes?

a. $C_{20}H_{42}$
b. $C_{20}H_{40}$
c. $C_{18}H_{38}$

Answer 8

(b) only.

(b) has the formula C_nH_{2n}, so may be an alkene. Both (a) and (c) have C_nH_{2n+2} formulas, so must be alkanes.

Question 9

Which of these alkenes are cyclic?

a. $C_{20}H_{40}$
b. $C_{20}H_{38}$
c. $C_{10}H_{18}$

Answer 9

(b) and (c) only.

These have C_nH_{2n-2} formulas. Compound (a) has the formula C_nH_{2n}, so must be an acyclic alkene.

Question 10

Name these alkenes:

a. $CH_3CH=CHCH_3$

b. 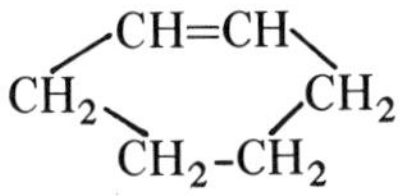
```
       CH=CH
      /     \
   CH₂       CH₂
      \     /
      CH₂-CH₂
```
(cyclohexene ring with one double bond)

c. $CH_3CH_2CH_2CH=CHCH_2CH_3$

d. $CH_3CH_2CH_2CH_2CH_2CH=CHCH_3$

e. $CH_2=CHCH_3$

Answer 10

a. 2-butene

b. cyclohexene

c. 3-heptene

d. 2-octene

e. propene

Using Table 20.1, we change -ane endings to -ene endings. A number is not needed to show the location of the double bonds in (b) or (e) because no confusion results from its omission: the identity of the compound is the same regardless of where the double bond is placed in the drawing.

Question 11

Show structures for these alkenes:

a. cycloheptene

b. 3-hexene

c. 1-nonene

Answer 11

a.
```
       CH₂—CH₂
      /        \
   CH₂          CH₂
   |           /
   CH₂        
      \      
       CH=CH
```

b. $CH_3CH_2CH=CHCH_2CH_3$

c. $CH_2=CH(CH_2)_6CH_3$ or $CH_3(CH_2)_6CH=CH_2$

Question 12

a. How many *sigma* bonds are there in a propene molecule?
b. How many *pi* bonds are there?

Answer 12

a. 8 *sigma* bonds

b. 1 *pi* bond

All bonds are *sigma* except the single *pi* bond involved in the double bond.

Statement 3

Alkynes are aliphatic hydrocarbons containing one or more **triple bonds**, each triple bond consisting of one **sigma** and two **pi** bonds. Alkynes are even more highly unsaturated than the alkenes, as can be seen from their formulas. The general formulas for alkynes and cycloalkynes are:

open chain alkynes	C_nH_{2n-2}
cycloalkynes	C_nH_{2n-4}

Nomenclature for the alkynes follows the same approach taken for alkenes. The -ane endings in Table 20.1 are replaced with -yne endings, and numerical prefixes are used when needed to locate the position of the triple bond.

Examples

$CH\equiv CCH_2CH_2CH_3$ $CH_3C\equiv CCH_2CH_3$

1-pentyne   2-pentyne

Note: Because of the linear geometry around the -C≡C- system, cycloalkynes must have at least 8 carbons in order to be stable.

Question 13

Which of these hydrocarbons may be alkynes?

a. C_4H_6

b. $C_{11}H_{20}$

c. $C_{11}H_{22}$

Answer 13

(a) and (b) only.

These have C_nH_{2n-2} formulas. Compound (c) has a C_nH_{2n} formula, so is either an alkane or cycloalkane.

Question 14

Name these alkynes.

a. $CH{\equiv}C(CH_2)_5CH_3$

b. $CH_3CH_2CH_2C{\equiv}C(CH_2)_4CH_3$

c. $CH{\equiv}CCH_3$

d. $CH{\equiv}CH$

Answer 14

a. 1-octyne

b. 4-decyne

c. propyne

d. ethyne

These are the "official" names (see Statement 10) for these alkynes. More commonly, however, compound (d) is called *acetylene* rather than ethyne, and the simple alkynes are named as derivatives of acetylene; for example, methylacetylene rather than propyne.

Question 15

Show structures for these alkynes.

a. 2-nonyne

b. 1-butyne

c. 5-dodecyne

PART 20—ORGANIC CHEMICALS

Answer 15

a. $CH_3C{\equiv}C(CH_2)_5CH_3$

b. $CH{\equiv}CCH_2CH_3$

c. $CH_3(CH_2)_3C{\equiv}C(CH_2)_5CH_3$

Question 16

a. How many *sigma* bonds are there in a propyne molecule?

b. How many *pi* bonds are there?

Answer 16

a. 6 *sigma* bonds

b. 2 *pi* bonds

$$H-\underset{H}{\overset{H}{\underset{|}{\overset{|}{C}}}}-C{\equiv}C-H$$

Only two of the bonds involved in the C≡C bond are *pi* bonds. The other six are *sigma* bonds.

Statement 4

Aromatic hydrocarbons are cyclic unsaturated hydrocarbons of a very special type. For example, the structure of **benzene**, the most common aromatic system, may be drawn according to standard conventions in one of the following ways:

expanded structural formula   condensed formula   abbreviated formula

The last of these, the **abbreviated formula**, is by far the most commonly used in showing cyclic structures of every type:

benzene
(aromatic)

cyclohexane
(aliphatic)

Because typical aromatic hydrocarbons are depicted with **alternating single and double bonds** by this method, we would expect them to react in much the same way as the cycloalkenes; but they do not! They are much **more stable** and **less reactive** than they appear to be in the preceding representations. In fact, all the bonds in benzene have the same length and strength, indicating that they are all alike: somewhere between a single and double bond. According to the valence bond approach to bonding, this is owing to **resonance**, which is depicted for benzene in the following way:

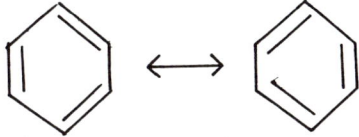

meaning that the true picture of benzene is a "hybrid" of the two drawings, and that the properties of benzene are somewhere between those predicted for the two pictures shown.

In molecular orbital theory, the same facts are accounted for by recognizing that all of the ring carbons have trigonal planar geometry, with six electrons located in **pi molecular orbitals** that involve all six of the carbons. In other words, the six electrons not being used in forming **sigma** C–C or C–H bonds are "smeared out" evenly over the molecule. Thus, we now indicate this delocalized **pi** bonding by a new abbreviation:

We may recognize whether a hydrocarbon is aromatic by use of the **Hückel rule**: Aromatic cyclic compounds will have **4n + 2 electrons in pi orbitals**, where n is any whole number. For any simple cyclic hydrocarbon system, whether "fused" (see naphthalene below) or isolated; when we draw the abbreviated structure in the old way, showing bonds as being either single or double: # π electrons = 2 × # double bonds

Example

π e⁻ = 2 × # double bonds

= 2 × 5

= 10

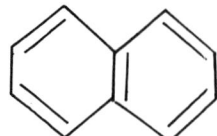

Testing: $4n + 2 = 10$

$n = 8/4$

$= 2$ which is a whole number.

Thus, naphthalene obeys the Hückel rule and **is aromatic**:

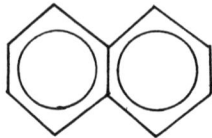

Question 17

What kind of overall geometry does benzene have?

Answer 17

a planar ring

All six carbons are sp^2 hybridized with trigonal planar geometry and with remaining p orbitals combining all around the ring to give the π molecular orbitals. Planarity is a feature shared by aromatic ring systems.

Question 18

How many *sigma* bonds are there in benzene?

Answer 18

12 *sigma* bonds

All six carbons are bonded to one H by a *sigma* bond; each C in the ring uses one *sigma* bond in bonding to the next ring C.

Question 19

How many electrons are located in *pi* orbitals in these cyclic hydrocarbons?

a. 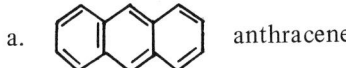 anthracene

b. cyclopentadiene

c. cyclooctatetraene

Answer 19

a. 14 *pi* electrons

b. 4 *pi* electrons

c. 8 *pi* electrons

OEIOU: #π e$^-$ = 2 × #double bonds

Question 20

Which of the hydrocarbons in Question 19 is aromatic?

Answer 20

Only (a), anthracene.

In (a): $4n + 2 = 14$
$n = 3$

In (b): $4n + 2 = 4$
$n = \frac{1}{2}$

In (c): $4n + 2 = 8$
$n = \frac{3}{2}$

Only in (a), then, does n have a whole-number value, so only anthracene obeys the Hückel rule. Note that cyclooctatetraene does have alternating single and double bonds, but is *not* aromatic (nor is it planar).

Question 21

Which of these cyclic hydrocarbons is aromatic?

a. (cycloheptatriene)

b. (phenanthrene)

c. (phenalene)

Answer 21

(b) only, with 14 π electrons.

Statement 5

Isomers are compounds that have the same numbers of **atoms** of each kind, but have different physical and/or chemical **properties**. These differences exist because the atoms and/or bonds are arranged differently in different isomers. Isomers may be classified broadly as follows:

structural isomers, which differ in the sequence or type of bonding of atoms to one another; or

stereoisomers, which differ in configuration about an atom (that is, the orientation in space of the atoms bonded to a central atom).

The acyclic alkanes shown earlier are all **straight-chain** compounds, with carbons connected one after another in a line. They may have isomers which are **branched-chain** compounds, with other carbons branching off the main carbon "backbone" of the alkane.

Example $CH_3CH_2CH_2CH_2CH_3$ $CH_3-\overset{\overset{\displaystyle CH_3}{|}}{CH}-CH_2CH_3$

n-pentane
(a straight-chain alkane)

isopentane
(a branched-chain alkane)

To distinguish among the possible alkane isomers, chemists often use the **prefixes** shown in Table 20.2.

As we shall see, other atoms may be substituted for a hydrogen on an alkane. To distinguish among several possible points of substitution, we give different names to the **alkyl groups** that result from the removal of different hydrogens. These are also shown in Table 20.2 for the first four alkanes.

Example CH_3CH_3 CH_3CH_2-
           ethane   ethyl
           (an alkane)   (an alkyl group)

TABLE 20.2 ISOMERIC ALKANES AND ALKYL GROUPS

Alkane	Name*	Alkyl Group	Name*
CH_4	methane	CH_3-	methyl
CH_3CH_3	ethane	CH_3CH_2-	ethyl
$CH_3CH_2CH_3$	propane	$CH_3CH_2CH_2-$	n-propyl
		$CH_3\underset{\vert}{C}HCH_3$	isopropyl
$CH_3CH_2CH_2CH_3$	n-butane	$CH_3CH_2CH_2CH_2-$	n-butyl
		$CH_3CH_2\underset{\vert}{C}HCH_3$	sec-butyl
$CH_3\underset{\underset{CH_3}{\vert}}{C}HCH_3$	isobutane	$CH_3\underset{\underset{CH_3}{\vert}}{C}HCH_2-$	isobutyl
		$CH_3\underset{\underset{CH_3}{\vert}}{C}CH_3$	t-butyl

*n- = normal, sec- = secondary, t- = tertiary

Question 22

Which of these are structural isomers of 1-hexene?

a. $CH_3CH=CHCH_2CH_2CH_3$

b. cyclohexene

c. $CH_3CH_2CH=CHCH_2CH_3$

d. $CH_2=\underset{\underset{CH_3}{\vert}}{C}-CH_2CH_2CH_3$

Answer 22

a, c, and d.

These compounds differ from 1-hexene only in the sequence or type of bonding of the carbons to one another. Cyclohexene has fewer hydrogens than any of these, so cannot be an isomer.

Question 23

Draw all the structural isomers of pentane.

Answer 23

$CH_3CH_2CH_2CH_2CH_3$ (n-pentane)

$CH_3\underset{\underset{CH_3}{|}}{C}HCH_2CH_3$ (isopentane)

$CH_3\underset{\underset{CH_3}{|}}{\overset{\overset{CH_3}{|}}{C}}CH_3$ (neopentane)

Notice that still another prefix, neo-, had to be introduced for C_5H_{12}, which was not needed for C_4H_{10}.

Question 24

Draw all the possible alkyl groups which could result from removing one H from n-pentane.

Answer 24

$CH_3CH_2CH_2CH_2CH_2-$

$CH_3CH_2CH_2\underset{|}{C}HCH_3$

$CH_3CH_2\underset{|}{C}HCH_2CH_3$

Question 25

Draw all the structural alkene isomers of 2-butene.

Answer 25

$CH_2=CHCH_2CH_3$

$CH_2=\underset{\underset{CH_3}{|}}{C}-CH_3$

Question 26

Which of these are structural isomers of 2-heptene?

a. CH$_3$CH—CH=CHCH$_2$CH$_3$
    |
    CH$_3$

b. cycloheptane

c. CH$_3$—CH—CH=C—CH$_3$
       | |
       CH$_3$ CH$_3$

Answer 26

All of them.

Note that even the saturated cycloheptane is an isomer; its formula, like that of (a), (b), and 2-heptene, is C$_7$H$_{14}$.

Statement 6

Geometric isomers exist for many **alkenes** because the **pi** bond holds groups bonded to the carbons rigidly in place on one side of the molecule. The position of groups (on either the same side or opposite sides of the alkene) is indicated by the prefix *cis-* or *trans-* on the name of the compound.

Example

butane	cis-2-butene trans-2-butene	2-butyne
Tetrahedral geometry about C.	Trigonal planar geometry about C.	Linear geometry about C.
Free rotation about C-C at room temperature.	Rigid π bond, so CH$_3$– groups held on **same** side or opposite sides.	All groups lined up, so molecule has no opposite "sides."
No geometric isomers possible.	Geometric isomers possible.	No geometric isomers possible.

(Structures: butane H$_3$C–CH$_2$–CH$_2$–CH$_3$ with tetrahedral C; cis-2-butene with both CH$_3$ on same side; trans-2-butene with CH$_3$ on opposite sides; 2-butyne CH$_3$–C≡C–CH$_3$)

Question 27

Is geometric isomerism a type of structural or Stereoisomerism?

Answer 27

Stereoisomerism.

In geometric isomers it is the orientation in space (configuration) that differs, not the order of bonding. (See Statement 5.)

Question 28

Which of these hydrocarbons may exist in *cis*- and *trans*- forms?

a. $CH_2=CH-CH_2CH_2CH_2CH_3$

b. $CH_3CH=CHCH_2CH_2CH_3$

c.
$$\begin{array}{c} CH_3 \\ \diagdown \\ CH_3 \end{array} CH-CH \begin{array}{c} CH_3 \\ \diagup \\ CH_3 \end{array}$$

d. $CH_3CH_2C\equiv C-CH_2CH_3$

Answer 28

(b) only.

In (a), 1-hexene, the same compound results regardless of which H on the first C is on the same side as the H on its neighboring C. In (c) there is free rotation about all C–C bonds, so the H atoms are not held rigidly in place on one side of the molecule. In (d) the *sp*-hybridized carbons give a linear geometry and allow for only one group to be bonded to each of the triple-bonded carbons.

Question 29

Draw the geometric isomers for 2-pentene and name each.

Answer 29

$$\begin{array}{c} H \\ \diagdown \\ CH_3 \end{array} C=C \begin{array}{c} H \\ \diagup \\ CH_2CH_3 \end{array} \quad \textit{cis}\text{-2-pentene}$$

$$\begin{array}{c} H \\ \diagdown \\ CH_3 \end{array} C=C \begin{array}{c} CH_2CH_3 \\ \diagup \\ H \end{array} \quad \textit{trans}\text{-2-pentene}$$

Statement 7

Benzene, like other hydrocarbons, may have other atoms (or groups) substituted for a hydrogen on the C-skeleton. When a single alkyl group is substituted for hydrogen, we name the resulting compound by placing the name of the alkyl group before the word "benzene."

Example: C$_6$H$_5$—CH$_2$CH$_3$ ethylbenzene

When **two groups** are substituted on the ring, the possible isomers are commonly distinguished from one another by using the prefixes *ortho-*, *meta-*, or *para-* (or simply *o-*, *m-*, *p-*).

Examples:

o-diethylbenzene   *m*-diethylbenzene   *p*-diethylbenzene

When **more than two groups** are substituted on the ring, the ring positions are numbered (giving the lowest possible numbers to carbons with substituents) and these numbers used as prefixes to distinguish among isomers. This numbering approach may also be used when only two substituents appear on the ring—indeed, this is the "official" system (see Statement 10)—but the use of *o-*, *m-*, and *p-* is more common.

Question 30

Name these compounds.

a. C$_6$H$_5$—CH(CH$_3$)—CH$_3$

b. C$_6$H$_5$—CH(CH$_3$)—CH$_2$CH$_3$

c. C$_6$H$_5$—CH$_2$—CH(CH$_3$)—CH$_3$

Answer 30

a. isopropylbenzene

b. *sec*-butylbenzene

c. isobutylbenzene

Question 31

Name these isomers.

a. 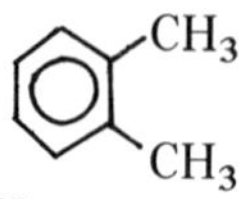

b. 

c. 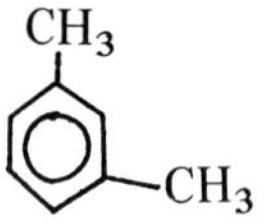

Answer 31

a. *o*-dimethylbenzene
(or 1,2-dimethylbenzene)

b. *p*-dimethylbenzene
(or 1,4-dimethylbenzene)

c. *m*-dimethylbenzene
(or 1,3-dimethylbenzene)

Question 32

Name these compounds.

a. 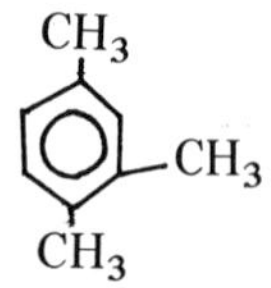

b. 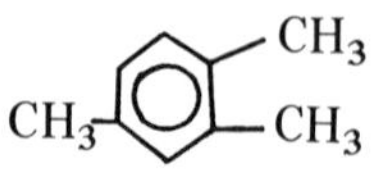

c. 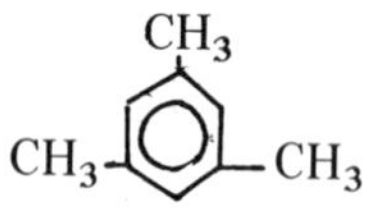

Answer 32

a. 1,2,4-trimethylbenzene

b. 1,2,4-trimethylbenzene

c. 1,3,5-trimethylbenzene
a and b are the same.

Statement 8

We may think of a typical organic compound as consisting of a carbon skeleton covered with a relatively inert hydrogen "skin" and possessing localized sites of reactivity—the **functional groups** that give organic compounds their chemical personalities.

The alkanes consist solely of C and H atoms, which differ little in electronegativity and are held together totally by **sigma** bonds. Thus they tend to be relatively inert. Alkenes, alkynes, and aromatic hydrocarbons, on the other hand, have regions of higher, more exposed electron density—the **pi** bonds—and show much greater chemical reactivity at these sites.

In other families of compounds a hydrogen has been replaced by an atom or group of atoms of some other kind. Because of the difference in attraction for electrons of this

TABLE 20.3 SOME TYPES OF SIMPLE ORGANIC COMPOUNDS

Functional Group	General Formula of Compound*	Type of Compound
$-X$ (F,Cl,Br,I)	$R-X$	halide
$-OH$	$R-OH$	alcohol
$-O-$	$R-O-R$	ether
$-NH_2$	$R-NH_2$	amine
$\overset{O}{\underset{\|\|}{-C-H}}$	$\overset{O}{\underset{\|\|}{R-C-H}}$	aldehyde
$\overset{O}{\underset{\|\|}{-C-}}$	$\overset{O}{\underset{\|\|}{R-C-R}}$	ketone
$\overset{O}{\underset{\|\|}{-C-OH}}$	$\overset{O}{\underset{\|\|}{R-C-OH}}$	acid
$\overset{O}{\underset{\|\|}{-C-O-}}$	$\overset{O}{\underset{\|\|}{R-C-O-R}}$	ester
$\overset{O}{\underset{\|\|}{-C-NH_2}}$	$\overset{O}{\underset{\|\|}{R-C-NH_2}}$	amide

*R represents any alkyl or aryl group.

functional group as compared to the C to which it is bonded, a **polar bond** may deplete (or more rarely, build up) the electron density on the carbon:

$$\overset{\delta^+}{\underset{}{>}}C \overset{\delta^-}{-} X$$

The presence of lone-pair electrons on the functional group may also provide a site of reactivity in the molecule. In every case, a functional group owes its reactivity to the build-up or depletion of electron density at its location in the molecule.

An important goal for the student at the introductory level of organic chemistry should be the recognition of the various families of organic compounds. (See Table 20.3.)

Question 33

What type of compound is each of these examples?

a. $CH_3-\underset{\underset{O}{\|}}{C}-OH$

b. $CH_3-\underset{\underset{O}{\|}}{C}-OCH_2CH_3$

c. ⟨○⟩−Cl

d. $\underset{CH_3}{\overset{CH_3}{>}}CH-O-CH_3$

Answer 33

a. acid

b. ester

c. halide

d. ether

Question 34

What family of organic compounds is represented by these examples?

a. $CH_3CH_2CH_2-\underset{\underset{}{\overset{O}{\|}}}{C}-CH_3$

b.

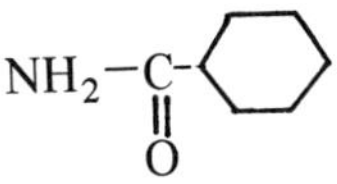

NH$_2$–C(=O)–C$_6$H$_{11}$

c. CH$_3$CH$_2$–NH$_2$

d. 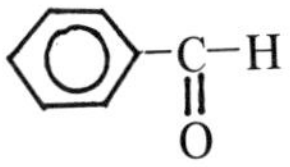 C$_6$H$_5$–C(=O)–H

Answer 34
a. ketone
b. amide
c. amine
d. aldehyde

Question 35

What type of compound is each of these examples?

a. 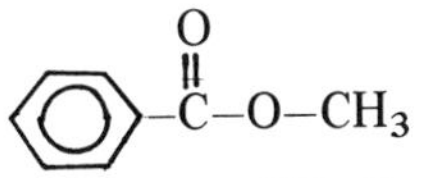 C$_6$H$_5$–C(=O)–O–CH$_3$

b. CH$_3$CH$_2$CH$_2$CH(CH$_3$)–C(=O)–O–CH$_2$CH$_3$

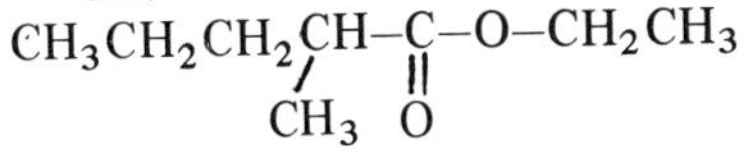

c. CH$_3$–CH(CH$_3$)–NH$_2$

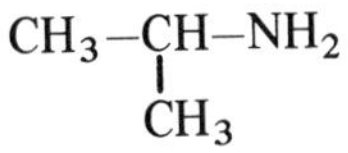

d. 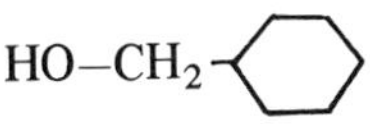 HO–CH$_2$–C$_6$H$_{11}$

Answer 35
a. ester
b. ester
c. amine
d. alcohol

Question 36

Which of these types of compounds contain *pi* bonds?

a. ketones

b. aldehydes

c. esters

d. ethers

Answer 36

All except (d), ethers.

Whether a double bond is between carbons, C=C, or between carbon and oxygen, C=O, one of the two bonds is a *pi* bond, the other a *sigma*.

Statement 9

Common names are generally used for the simpler organic compounds, especially the alcohols, halides, and amines. The name of the hydrocarbon group (Table 20.2) is given first, followed by the name of the compound type (Table 20.3). (See Statements 10 and 11 for discussion of the "official" system of nomenclature.)

Question 37

What are the common names of these compounds?

a. $CH_3-CH-OH$
         $|$
         CH_3

b. $CH_3CH_2CH_2-Br$

c. $CH_3CH_2-NH_2$

Answer 37

a. isopropyl alcohol

b. *n*-propyl bromide

c. ethylamine

Note: In naming the amines, the name is made into a single word without a space between the group name and the name of the compound type. Spaces are retained for the alcohols and halides.

Question 38

Give the structures corresponding to these common names.

a. isobutyl iodide

b. *t*-butyl alcohol

c. *sec*-butylamine

Answer 38

a. $CH_3-\underset{\underset{\displaystyle CH_3}{|}}{CH}-CH_2-I$

b. $CH_3-\underset{\underset{\displaystyle CH_3}{|}}{\overset{\overset{\displaystyle CH_3}{|}}{C}}-OH$

c. $CH_3\underset{\underset{\displaystyle NH_2}{|}}{CH}-CH_2CH_3$

Statement 10

The common names are feasible only for the more uncomplicated organic compounds. Molecules containing very many atoms may have literally hundreds (or thousands) of isomers, so that we quickly run out of distinguishing prefixes.

IUPAC* names provide distinct, systematic descriptions of organic compounds. These "official" names often suffer from an unwieldiness that robs them of their usefulness in oral communication. They are essential, however, in writing about moderately complex organic structures.

The general approach taken in the **IUPAC nomenclature of aliphatic compounds** is outlined below:

(1) The **longest continuous chain** of C atoms is considered to be the "parent" hydrocarbon (Table 20.1), and any branching alkyl or functional groups attached to it are considered to be substituents of that parent.

(2) Number the parent chain so as to give the **lowest possible set of numbers** to the carbons bearing alkyl or functional groups.

(3) Name the compound by citing the number of the carbon to which a group is attached, then the name of the group (Table 20.2 and Table 20.4). Cite different attached groups in alphabetic order. (The following example shows the required punctuation and the use of prefixes when more than one group of a given kind is present.)

(4) End the name with the name of the parent hydrocarbon.

---

*International Union of Pure and Applied Chemistry

PART 20—ORGANIC CHEMICALS

TABLE 20.4. SOME COMMON FUNCTIONAL GROUP NAMES USED IN IUPAC NOMENCLATURE

—F	fluoro
—Cl	chloro
—Br	bromo
—I	iodo
—NH_2	amino
—NO_2	nitro

Example

To name this compound

$$CH_3-CH_2-\underset{\underset{CH_3}{\underset{|}{CH_2}}}{\overset{}{\underset{|}{CH}}}-CH_2-\underset{\underset{Cl}{|}}{\overset{\overset{Cl}{|}}{C}}-CH_3$$

Step (1):   C–C–C–C–C–C   "hexane" chain is longest
                   | continuous C-chain
                   C
                   |
                   C

Step (2):
$$\underset{\underset{CH_3}{\underset{|}{CH_2}}}{\overset{}{\underset{6\ 5\ 4\ 3}{C-C-C-C}}}\underset{\underset{}{\underset{2}{|}}\ \underset{1}{}}{\overset{\overset{Cl}{|}}{-C-C}}$$
Cl below position 2

lowest possible set of numbers for C-atoms with groups attached

Step (3):   2,2-dichloro-4-ethyl   names of groups with numbers showing points of attachment

Step (4):   2,2-dichloro-4-ethylhexane   **final name**

Question 39

Name this compound according to the IUPAC method.

$$CH_3CH_2\underset{\underset{F}{|}}{CH}CH_2CH_2CH_3$$

Answer 39

3-fluoroheptane

Question 40

Name this compound, using the IUPAC method.

$$\begin{array}{c} CH_3 \quad\quad Br \\ \searrow| \\ CH-CH-CH-NH_2 \\ \nearrow| \\ CH_3 \quad\quad\quad CH_2CH_3 \end{array}$$

Answer 40

4-amino-3-bromo-2-methylhexane

(*not* 3-amino-4-bromo-5-methylhexane); 4-3-2 is a lower set of numbers (total of 9) than 3-4-5 (total of 12).

Question 41

Give the IUPAC name for this compound:

$$\begin{array}{c} CH_3 \\ | \\ I-(CH_2)_3-C-CH_2-CH-CH_3 \\ || \\ CH_3CH_3 \end{array}$$

Answer 41

1-iodo-4,4,6-trimethylheptane

Question 42

Name this compound.

$$\begin{array}{c} CH_2CH_3 \\ | \\ CH_3-CH-CH-CH-NO_2 \\ || \\ ClCH_2-CH_2-CH_2-CH_3 \end{array}$$

Answer 42

4-chloro-3-methyl-5-nitrononane

Question 43

Draw the structures corresponding to these names.

a. 1-chloro-2,2,4-trimethylpentane
b. 3-chloro-5-ethyl-2-fluorooctane

Answer 43

a.
$$\text{Cl-CH}_2-\underset{\underset{CH_3}{|}}{\overset{\overset{CH_3}{|}}{C}}-\text{CH}_2-\underset{}{\overset{\overset{CH_3}{|}}{CH}}-\text{CH}_3$$

b.
$$\text{CH}_3\text{CH}_2-\underset{\underset{F}{|}}{\overset{\overset{Cl}{|}}{C}}-\text{CH}_2-\underset{\underset{CH_2CH_3}{|}}{CH}-\text{CH}_2\text{CH}_2\text{CH}_3$$

Statement 11

When certain key functional groups are present, the IUPAC considers these so important that the parent chain must be chosen so that it contains these groups (or bears them as substituents). The name then will have, not -ane, but some other appropriate ending which denotes the functional group. Some notable examples:

group	parent ending	example	IUPAC name
C=C	-ene	$\text{CH}_2\text{=C-CH}_2\text{CH}_2\text{CH}_3$ with CH_2CH_3 branch	2-ethyl-1-pentene
$-\text{C}\equiv\text{C}-$	-yne	$\text{CH}\equiv\text{C-CH-CH}_2\text{CH}_3$ with CH_2CH_3 branch	3-ethyl-1-pentyne
$-\text{OH}$	-ol	$\text{CH}_3-\text{CH-CH-CH}_2\text{CH}_3$ with OH and CH_2CH_3	3-ethyl-2-pentanol
$-\text{CH=O}$	-al	$\text{CH}_3-\text{CH-CH-CH}_2\text{CH}_3$ with CH=O and CH_2CH_3	3-ethyl-2-methylpentanal
$\overset{O}{\underset{}{\overset{\|}{-C}}}-\text{OH}$	-oic acid	$\text{CH}_3-\text{CH-CH-CH}_2\text{CH}_3$ with C(=O)-OH and CH_2CH_3	3-ethyl-2-methylpentanoic acid

Question 44

Name these compounds, using the IUPAC method.

a. $CH_3\underset{\underset{OH}{|}}{CH}CH_3$

b. $CH_3CH_2CH=O$

c. $CH_3CH_2-\underset{\underset{O}{||}}{C}-OH$

Answer 44

a. 2-propanol
   (common name: isopropyl alcohol)

b. propanal
   (common name: propionaldehyde)

c. propanoic acid
   (common name: propionic acid)

Question 45

Name these compounds, using the IUPAC method.

a. $CH_3CH_2CH_2-\underset{\underset{CH=CH_2}{|}}{CH}-CH_2CH_2CH_3$

b. $CH\equiv C-\underset{\underset{CH_2CH_2CH_3}{|}}{CH}-CH_2CH_2CH_3$

Answer 45

a. 3-propyl-1-hexene

b. 3-propyl-1-hexyne

Question 46

What are the IUPAC names for these alcohols?

a. $CH_3-\underset{\underset{CH_3}{|}}{CH}-CH_2-OH$

b. $CH_3-\underset{\underset{OH}{|}}{CH}-CH_2CH_3$

Answer 46

a. 2-methyl-1-propanol
   (common name: isobutyl alcohol)

b. 2-butanol
   (common name: sec-butyl alcohol)

Statement 12

Electrophilic substitutions are organic reactions which result in the displacement of a group on the molecule by an electrophilic ("electron-loving") reagent. These are likely reactions for compounds in which an alkyl group (R-) or an aryl group (Ar-) is attached to an element **less electronegative** than C:

$$\underset{}{\overset{\delta^-\delta^+}{>C-Z}}$$

Some common types of electrophilic substitution

aliphatic $R-H \xrightarrow{X_2, \text{ light}} R-X$ (X = Cl or Br)

$R-Li \xrightarrow{X_2} R-X$

aromatic $Ar-H \xrightarrow{X_2, \text{ Fe}} Ar-X$

$Ar-H \xrightarrow[H_2SO_4]{HNO_3} Ar-NO_2$

$Ar-H \xrightarrow[SO_3]{H_2SO_4} Ar-SO_3H$

Notice that some of the reactions require specific **catalysts**. Notice, too, the "shorthand" approach typically taken by organic chemists in writing equations, with reactants and catalysts written over and under the arrow and the organic compounds used as reactants and products shown on the left and right of the arrow.

Question 47

Predict the organic products of these reactions.

a. $CH_3CH_3 \xrightarrow{Cl_2, \text{ light}}$

b. benzene $\xrightarrow{Cl_2, \text{ Fe}}$

Answer 47

CH$_3$CH$_2$—Cl ethyl chloride (IUPAC: chloroethane)

b. Cl—⟨◯⟩ chlorobenzene
(common name: phenyl chloride)

Question 48

Predict the organic products of these reactions.

a. ⟨◯⟩ $\xrightarrow{\text{HNO}_3}{\text{H}_2\text{SO}_4}$

b. ⟨◯⟩ $\xrightarrow{\text{H}_2\text{SO}_4}{\text{SO}_3}$

Answer 48

a. ⟨◯⟩—NO$_2$ nitrobenzene

b. ⟨◯⟩—SO$_3$H benzenesulfonic acid

Statement 13

In **nucleophilic substitution** a reagent attacks an electron-deficient carbon atom, resulting in the displacement of some other group bonded there. This reaction is likely when an R- (or Ar-, though this is much rarer) group is attached to a group that is **more electronegative** than C:

$$\underset{\delta^+}{>\!\!C}\!-\!\underset{\delta^-}{X}$$

Some Common Types of Nucleophilic Substitution

R—X $\xrightarrow{\text{OH}^-}$ R—OH (X = halogen)

R—X $\xrightarrow{\text{H}_2\text{O}}$ R—OH

R—X $\xrightarrow{\text{NH}_2^-}$ R—NH$_2$

R—X $\xrightarrow{\text{OR}'^-}$ R—O—R'

R—OH $\xrightarrow{\text{HX}}$ R—X

Question 49

Predict the organic product resulting from each of these reactions.

a. $CH_3CH_2-Br \xrightarrow{KNH_2}$

b. $CH_3CH-OH \xrightarrow{HBr}$
       $|$
       CH_3

c. $CH_3\underset{\underset{CH_3}{|}}{\overset{\overset{CH_3}{|}}{C}}-Cl \xrightarrow{H_2O}$

Answer 49

a. $CH_3CH_2-NH_2$ ethylamine

b. $CH_3-CH-Br$ isopropyl bromide
       $|$
       CH_3

c. $CH_3\underset{\underset{CH_3}{|}}{\overset{\overset{CH_3}{|}}{C}}-OH$ t-butyl alcohol

---

Question 50

Suggest a reagent which could be used to convert 2-bromobutane into:

a. $CH_3-CH-CH_2CH_3$
        $|$
        OCH_3

b. $CH_3-CH-CH_2CH_3$
        $|$
        OH

Answer 50

a. OCH_3^-

(actually $NaOCH_3$ or $KOCH_3$, which are analogous to NaOH and KOH)

b. OH^-

(actually NaOH or KOH)

Statement 14

Electrophilic additions may take place when a molecule has a double or triple bond. They are major reactions of the **alkenes** and **alkynes** but not of benzene and other aromatic hydrocarbons, because of resonance stabilization. When more than one product may result from the addition of a compound, H-X, to an alkene or alkyne, the **major product** is usually the one with the incoming H attached to the C with the greater number of hydrogens already bonded to it:

$$\begin{array}{c} R \\ \diagdown \\ C=C \\ \diagup \\ R \end{array} \begin{array}{c} H \\ \diagup \\ \\ \diagdown \\ H \end{array} \xrightarrow{\text{H-X}} \begin{array}{c} X\ H \\ |\ \ | \\ R-C-C-H \\ |\ \ | \\ R\ H \end{array}$$

Some Common Electrophilic Addition Reactions

$$\text{C=C} \xrightarrow{HX} \begin{array}{c} H\ X \\ |\ \ | \\ -C-C- \\ |\ \ | \end{array} \qquad (X = \text{halogen})$$

$$\text{C=C} \xrightarrow[\text{acid}]{H_2O} \begin{array}{c} |\ \ | \\ -C-C- \\ |\ \ | \\ H\ OH \end{array}$$

$$\text{C=C} \xrightarrow{X_2} \begin{array}{c} |\ \ | \\ -C-C- \\ |\ \ | \\ X\ X \end{array}$$

Question 51

Predict the resulting major organic product when these reagents are mixed with propene.

a. Br_2

b. HBr

c. H_2O/acid

Answer 51

a. $CH_3-\underset{Br}{CH}-\underset{Br}{CH_2}$ 1,2-dibromopropane

b. $CH_3-\underset{Br}{CH}-CH_3$ 2-bromopropane (or isopropyl bromide)

c. $CH_3-\underset{OH}{CH}-CH_3$ 2-propanol (or isopropyl alcohol)

Question 52

What reagents might be used to prepare these compounds from 1-butene?

a. 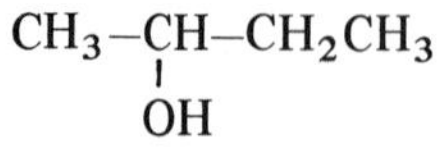 $CH_3-\underset{\underset{OH}{|}}{CH}-CH_2CH_3$

b. 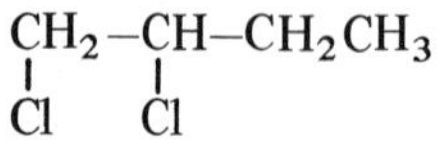 $\underset{\underset{Cl}{|}}{CH_2}-\underset{\underset{Cl}{|}}{CH}-CH_2CH_3$

c. 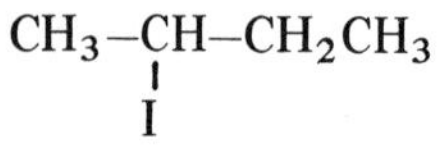 $CH_3-\underset{\underset{I}{|}}{CH}-CH_2CH_3$

Answer 52

a. H_2O/acid

b. Cl_2

c. HI

Question 53

Predict the major product of these reactions.

a. $CH\equiv CH + 2Br_2$

b. $CH\equiv CH + 2HBr$

Answer 53

a. 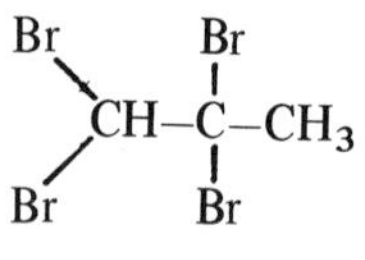 $\underset{Br}{\overset{Br}{\diagdown}}CH-\underset{\underset{Br}{|}}{\overset{\overset{Br}{|}}{C}}-CH_3$

b. 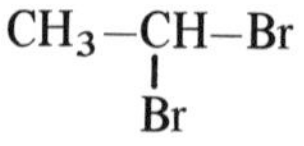 $CH_3-\underset{\underset{Br}{|}}{CH}-Br$

Statement 15

Nucleophilic additions are common reactions for **aldehydes** and **ketones**, in which the carbonyl group appears:

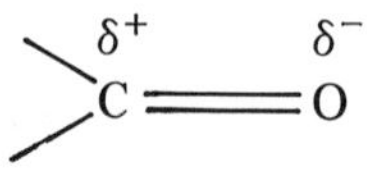

$$\overset{\delta+}{\underset{}{>}}C=\overset{\delta-}{O}$$

PART 20—ORGANIC CHEMICALS

Notice that the carbon now is somewhat positive, and so becomes attractive to nucleophiles (such as the CN^- ion, which is negative).

Some Common Types of Nucleophilic Addition

$$R-\underset{\substack{\|\\O}}{C}-R' \xrightarrow{HCN} R-\underset{\substack{|\\CN}}{\overset{\substack{OH\\|}}{C}}-R'$$

$$R-\underset{\substack{\|\\O}}{C}-R' \xrightarrow[\text{acid}]{NH_2-NH_2} R-\underset{\substack{|\\R'}}{C}=N-NH_2 + H_2O$$

$$R-\underset{\substack{\|\\O}}{C}-R' \xrightarrow[\text{acid}]{NH_2-OH} R-\underset{\substack{|\\R'}}{C}=N-OH + H_2O$$

R' as used above may represent hydrogen as well as alkyl or aryl groups. Note that in the last two reactions the process of addition is accompanied by loss of H_2O.

Question 54

Predict the organic products of these reactions.

a. $CH_3-CH=O + NH_2OH$/acid

b. $CH_3-CH=O + HCN$

c. $CH_3-\underset{\substack{\|\\O}}{C}-CH_3$

Answer 54

a. $CH_3CH=N-OH$

b. $CH_3-\underset{\substack{|\\CN}}{CH}-OH$

c. $CH_3-\underset{\substack{|\\OH}}{\overset{}{C}}\underset{\substack{|\\CN}}{}-CH_3$

Question 55

Predict the organic products of these reactions.

a. $CH_3CH_2CH=O \xrightarrow[\text{acid}]{NH_2OH}$

PART 20–ORGANIC CHEMICALS

b. $CH_3-\underset{\underset{O}{\|}}{C}-CH_2CH_3 \xrightarrow[\text{acid}]{NH_2NH_2}$

Answer 55

a. $CH_3CH_2CH=N-OH$

b. $CH_3-\underset{\underset{N-NH_2}{\|}}{C}-CH_2CH_3$

Question 56

What reagents are needed to obtain these products from butanal?

a. $HO-\underset{\underset{CN}{|}}{CH}-CH_2CH_2CH_3$

b. $CH_3CH_2CH_2CH=N-NH_2$

Answer 56

a. HCN

b. NH_2-NH_2/acid

Statement 16

In **elimination reactions** groups are removed from the molecule but are not replaced. The resulting compounds are always unsaturated. The initial attack on the molecule is usually an attack by a nucleophile. Notice that eliminations are the reverse of addition reactions.

When more than one product is possible, the major product will usually be the one with more R- groups substituted on the C=C system.

Some Common Types of Elimination Reaction

$\underset{\underset{|\ \ |}{|\ \ |}}{-C-C-}\overset{H\ OH}{} \xrightarrow[\text{heat}]{\text{acid}} \diagup C=C \diagdown$

$\underset{\underset{|\ \ |}{|\ \ |}}{-C-C-}\overset{H\ X}{} \xrightarrow[\text{heat}]{OH^-} \diagup C=C \diagdown \qquad (X = halogen)$

$$\underset{\substack{|\ |\\H\ X}}{\overset{\substack{H\ X\\|\ |}}{-C-C-}} \xrightarrow[\text{heat}]{2OH^-} -C\equiv C-$$

(Notice that the preceding two equations give an alternative route for the reaction between an alkyl halide and the OH^- ion to that shown in the nucleophilic substitution reaction. We shall not discuss the method for predicting which route wins out in such competing processes.)

Question 57

Predict the major elimination products of these reactions.

a. $CH_3-\underset{\underset{Br}{|}}{CH}-CH_3 + KOH/\text{heat}$

b. $CH_3-\underset{\underset{OH}{|}}{CH}-CH_3 + H_2SO_4/\text{heat}$

c. $CH_3CH_2CH_2-Br + KOH/\text{heat}$

Answer 57

a. $CH_3CH=CH_2$

b. $CH_3CH=CH_2$

c. $CH_3CH=CH_2$

Question 58

Predict the major elimination products of these reactions.

a. $CH_3-\underset{\underset{Cl}{|}}{CH}-CH_2CH_3 + KOH/\text{heat}$

b. $Cl-CH_2CH_2CH_2CH_3 \xrightarrow[\text{heat}]{KOH}$

c. $CH_3-\underset{\underset{Cl}{|}}{CH}-Cl + 2KOH/\text{heat}$

Answer 58

a. $CH_3CH=CHCH_3$

b. $CH_2=CH-CH_2CH_3$

c. $CH\equiv CH$

PART 20—ORGANIC CHEMICALS

Statement 17

Some other organic reactions that involve the combination of portions of two molecules are summarized below.

Some Miscellaneous "Combination" Reactions

reactants → product

$$R-OH + R'-\underset{\underset{O}{\|}}{C}-OH \xrightarrow{H^+} R'-\underset{\underset{O}{\|}}{C}-O-R$$
alcohol + acid → ester

$$R-NH_2 + R'-\underset{\underset{O}{\|}}{C}-OH \xrightarrow[H^+]{heat} R'-\underset{\underset{O}{\|}}{C}-NH-R \quad \text{(Also for R— = H)}$$
amide

$$2\ R-OH \xrightarrow{H^+} R-O-R$$
alcohol → ether

$$R-CH=O + R'-OH \xrightarrow{H^+} R-CH\underset{OR'}{\overset{OR'}{\diagdown}}$$
aldehyde + alcohol → acetal

Question 59

Predict the product of the reaction of ethanol with:

a. $CH_3-\underset{\underset{O}{\|}}{C}-OH/H^+$

b. CH_3CH_2-OH/H^+

c. $CH_3CH=O/H^+$

Answer 59

a. $CH_3CH_2-O-\underset{\underset{O}{\|}}{C}-CH_3$

b. $CH_3CH_2-O-CH_2CH_3$

  (In the lab this would entail simply treating ethanol itself with a source of H^+, such as H_2SO_4.)

c. $CH_3-\underset{\underset{O-CH_2CH_3}{|}}{CH}-O-CH_2CH_3$

Question 60

Predict the products of these reactions.

a. $CH_3CH_2CH_2-NH_2 + CH_3-\underset{\underset{O}{\|}}{C}-OH$ /heat

b. $CH_3CH_2CH_2-OH + CH_3-\underset{\underset{O}{\|}}{C}-OH/H^+$

Answer 60

a. $CH_3CH_2CH_2-NH-\underset{\underset{O}{\|}}{C}-CH_3$

b. $CH_3CH_2CH_2-O-\underset{\underset{O}{\|}}{C}-CH_3$

Question 61

What reagent would have to be reacted with isobutyl alcohol to give the following:

a. $CH_3CH_2-\underset{\underset{O}{\|}}{C}-O-CH_2CH_2\underset{\underset{CH_3}{|}}{C}HCH_3$

b. $CH_3\underset{\underset{CH_3}{|}}{C}HCH_2-O-CH_2\underset{\underset{CH_3}{|}}{C}HCH_3$

Answer 61

a. $CH_3CH_2CH_2-\underset{\underset{O}{\|}}{C}-OH/H^+$

b. Treat with a source of H^+ such as H_2SO_4.

Statement 18

Oxidation and reduction reactions are common for many types of organic compounds. The products of these processes depend to a great extent on the severity of the conditions under which they are carried out. With very strong oxidizing agents and high temperatures, for example, all of the C and H in an organic compound will be converted into CO_2 and H_2O. In the following summary we shall symbolize oxidation by [O] without specifying whether the specific agent was O_2, MnO_4^-, $Cr_2O_7^{2-}$, and so forth, and will assume that complete oxidation does **not** take place.

Some Common Types of Reduction Reactions

$$\diagup_{C=C}\diagdown \xrightarrow{H_2 \text{ cat.}} \begin{array}{c} | \ | \\ -C-C- \\ | \ | \\ H \ H \end{array} \qquad (\text{cat.} = \text{Pt, Ni, Pd, etc.})$$

$$-C\equiv C- \xrightarrow{2H_2 \text{ cat.}} \begin{array}{c} H \ H \\ | \ | \\ -C-C- \\ | \ | \\ H \ H \end{array}$$

$$R-CH=O \xrightarrow{H_2 \text{ cat.}} R-CH_2-OH$$

$$\begin{array}{c} R-C=O \\ | \\ R \end{array} \xrightarrow{H_2 \text{ cat.}} \begin{array}{c} R-CH-OH \\ | \\ R \end{array}$$

Some Common Types of Oxidation Reactions

$$R-CH_2-OH \xrightarrow{[O]} R-CH=O$$

$$\begin{array}{c} R-CH-OH \\ | \\ R \end{array} \xrightarrow{[O]} \begin{array}{c} R-C=O \\ | \\ R \end{array}$$

$$R-CH=O \xrightarrow{[O]} \begin{array}{c} R-C=O \\ | \\ OH \end{array}$$

Question 62

Predict the products obtained when these reductions take place.

a. $CH_3-CH=CH_2 \xrightarrow{H_2}{Ni}$

b. $CH_3-CH=O \xrightarrow{H_2}{Pt}$

c. $CH_3-C\equiv CH \xrightarrow{2H_2}{Pd}$

d. $CH_3-C\equiv CH \xrightarrow{H_2}{Pd}$

Answer 62

a. $CH_3-CH_2-CH_3$

b. CH_3-CH_2-OH

c. $CH_3-CH_2-CH_3$

d. $CH_3-CH=CH_2$

It is possible to control conditions so that either one mole of H_2 or two moles of H_2 will add to the $C{\equiv}C$ bond.

Question 63

Predict the products of the oxidation of:

a. 1-butanol

b. 2-butanol

c. butanal

Answer 63

a. $CH_3CH_2CH_2CH{=}O$

b. $CH_3CH_2-\underset{\underset{O}{\|}}{C}-CH_3$

c. $CH_3CH_2CH_2-\underset{\underset{O}{\|}}{C}-OH$

TESTING YOUR MASTERY—Part 20 Organic Chemistry

Question 64

Below are the formulas of some saturated hydrocarbons. Classify each as an open chain alkane or a cycloalkane.

a. $C_{11}H_{22}$

b. C_7H_{16}

c. C_6H_{14}

d. C_3H_6

Answer 64

a. cycloalkane

b. open chain alkane

c. open chain alkane

d. cycloalkane

Question 65

Below are the formulas of some acyclic, unsaturated hydrocarbons. Which are alkenes and which are alkynes?

a. $C_{11}H_{22}$

b. C_7H_{12}

c. C_6H_{12}

d. C_8H_{14}

Answer 65

a. alkene

b. alkyne

c. alkene

d. alkyne

Question 66

Name these straight-chain alkanes according to the IUPAC system:

a. C_5H_{12}

b. C_7H_{16}

c. $C_{10}H_{22}$

Answer 66

a. pentane

b. heptane

c. decane

To formulate common names, the prefix *n-* is added to the above names.

Question 67

Name these compounds, using the IUPAC method:

a. $CH_3CH_2\underset{\underset{CH_3}{|}}{CH}CH_2CH_3$

b. $CH_2{=}CH-\underset{\underset{CH_3}{|}}{CH}CH_2CH_3$

c. $CH_3CH_2-\underset{\underset{CH_3}{|}}{CH}-C{\equiv}CH$

Answer 67

a. 3-methylpentane

b. 3-methyl-1-pentene

c. 3-methyl-1-pentyne

Question 68

Give the "official" names for these compounds:

a.
$$\underset{CH_3CH_2\qquad H}{\overset{H\qquad CH_2CH_3}{C{=}C}}$$

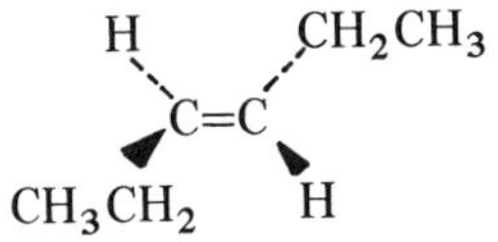

b. 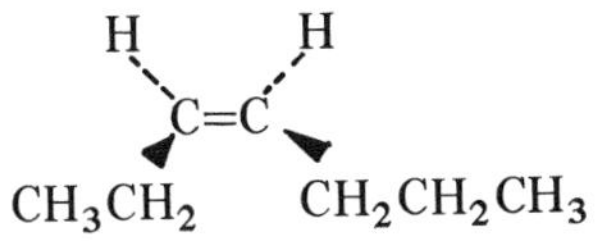
$$\underset{CH_3CH_2\quad CH_2CH_2CH_3}{\overset{H\qquad H}{C{=}C}}$$

c.
$$\underset{H\qquad CH_2CH_3}{\overset{H\qquad H}{C{=}C}}$$

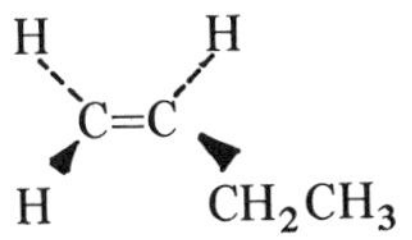

Answer 68

a. *trans*-3-hexene

b. *cis*-3-heptene

c. 1-butene

Question 69

Which of these pairs of compounds are structural isomers of each other?

a. $CH_3CH_2CH_2\underset{\underset{CH_3}{|}}{C}HCH_3$ and $CH_3\underset{\underset{CH_3}{|}}{C}H-\underset{\underset{CH_3}{|}}{C}HCH_3$

b. $CH_3CH_2-O-CH_2CH_3$ and $CH_3CH_2CH_2CH_2OH$

c. $CH_3CH_2CH=O$ and $CH_3-\underset{\underset{O}{||}}{C}-CH_3$

d. $CH_3CH_2CH_2CH_2CH_2CH_3$ and ⬡

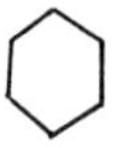

Answer 69

(a), (b), and (c).

Question 70

Write structures corresponding to these names:

a. isobutane

b. 2-bromo-5-ethyloctane

c. cyclopropane

d. *sec*-butyl alcohol

Answer 70

a. $CH_3-\underset{\underset{CH_3}{|}}{C}H-CH_3$

b. $CH_3-\underset{\underset{Br}{|}}{C}H-CH_2CH_2-\underset{\underset{CH_2CH_3}{|}}{C}H-CH_2CH_2CH_3$

c. 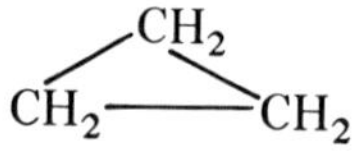 $CH_2\diagup{}^{CH_2}\diagdown{}_{CH_2}$ (cyclopropane ring)

d. $CH_3-\underset{\underset{OH}{|}}{C}H-CH_2CH_3$

Question 71

Name these compounds.

a. 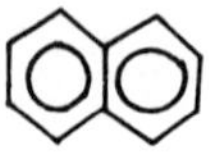 [naphthalene structure]

PART 20—ORGANIC CHEMICALS

b. 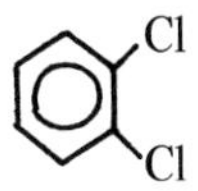 [structure: benzene ring with two Cl at adjacent positions]

c. 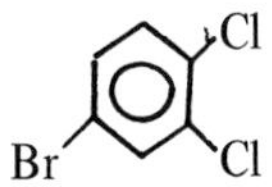 [structure: benzene ring with Br, Cl, Cl substituents]

Answer 71

a. naphthalene

b. o-dichlorobenzene (or 1,2-dichlorobenzene)

c. 4-bromo-1,2-dichlorobenzene

Question 72

What are the structures of these compounds?

a. isobutyl iodide

b. propanal

c. t-butylamine

Answer 72

a. CH_3CHCH_2-I
  $|$
  CH_3

b. $CH_3CH_2CH=O$

c. CH_3
  $|$
  CH_3C-NH_2
  $|$
  CH_3

Question 73

Name these compounds according to the IUPAC method:

a. $CH_3(CH_2)_3CH(CH_2)_3CH_3$
  $|$
  CH_2CH_2-OH

b. $CH_3(CH_2)_3CH-C\equiv C-CH_3$
  $|$
  $CH_2CH_2CH_2CH_3$

Answer 73

a. 3-butyl-1-heptanol
b. 4-butyl-2-octyne

Question 74

Which of these compounds are aromatic?

a.

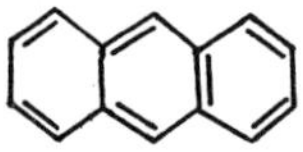

b.

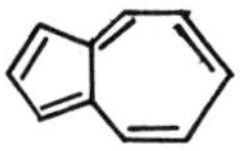

c.

Answer 74

(a) and (b)

Question 75

What types of compounds are these?

a. CH_3-NH_2

b. $CH_3-\underset{\underset{O}{\|}}{C}-NH_2$

c. $CH_3CH_2-O-CH_2CH_2CH_3$

Answer 75

a. amine
b. amide
c. ether

Question 76

Which of these compounds are likely to participate in electrophilic addition reactions?

a. cyclohexene

b. 3-hexyne

c. benzene

d. butanal

Answer 76

(a) and (b) only.

Alkenes and alkynes react easily in electrophilic addition.

Question 77

Which of these compounds are likely to undergo nucleophilic addition?

a. butanal

b. $CH_3-\underset{\underset{O}{\|}}{C}-CH_3$

c. ethyne

Answer 77

(a) and (b) only.

Aldehydes and ketones are most susceptible to nucleophilic addition.

Question 78

Predict the oxidation product from each of these compounds:

a. 2-butanol

b. 4-methyl-1-pentanal

c. 4-methyl-1-pentanol

Answer 78

a. $CH_3-\underset{\underset{O}{\|}}{C}-CH_2CH_3$

b. $CH_3CHCH_2CH_2-\underset{O}{\overset{\|}{C}}-OH$
        $|$
       CH_3

c. $CH_3CHCH_2CH_2CH=O$
       $|$
      CH_3

Question 79

Predict the major elimination product in these reactions.

a. $CH_3CH_2CH_2\underset{\underset{Br}{|}}{C}HCH_3 \xrightarrow{\text{KOH, heat}}$

b. $CH_3\underset{\underset{CH_3}{|}}{\overset{\overset{CH_3}{|}}{C}}-CH_2CH_2-OH \xrightarrow{\text{H}_2\text{SO}_4, \text{heat}}$

c. $CH_3\underset{Cl\ Cl}{\overset{\diagup\diagdown}{C}}-CH_2CH_3 \xrightarrow{\text{2NaOH, heat}}$

Answer 79

a. $CH_3CH_2CH=CHCH_3$

b. $CH_3\underset{\underset{CH_3}{|}}{\overset{\overset{CH_3}{|}}{C}}-CH=CH_2$

c. $CH_3C\equiv CCH_3$

Question 80

What would be the major products of these reactions?

a. [benzene ring] $\xrightarrow{\text{HNO}_3, \text{H}_2\text{SO}_4}$

b. [benzene ring] $\xrightarrow{\text{Br}_2, \text{Fe}}$

c. $CH_3CH_3 \xrightarrow{\text{Br}_2, \text{light}}$

Answer 80

a. 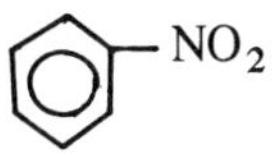

b. 

c. $CH_3CH_2\text{-}Br$

Question 81

Predict the major products of these reactions.

a. $CH_3CH_2CH{=}CH_2 \ + \ HI$

b. $CH_3C{=}CH_2 \ + \ H_2O/H^+$
$\quad\quad |$
$\quad\quad CH_3$

c. 2-methylpropene + Br_2

Answer 81

a. $CH_3CH_2CHCH_3$
$\quad\quad\quad\quad |$
$\quad\quad\quad\quad I$

$\quad\quad OH$
$\quad\quad |$
b. CH_3CCH_3
$\quad\quad |$
$\quad\quad CH_3$

$\quad\quad Br$
$\quad\quad |$
c. $CH_3CCH_2{-}Br$
$\quad\quad |$
$\quad\quad CH_3$

Question 82

Predict the major organic product when ethanal is treated with the following:

a. CH_3OH/H^+

b. NH_2OH/H^+

c. NH_2NH_2/H^+

d. H_2/Ni

Answer 82

a. $CH_3-\underset{\underset{CH_3}{|}}{\overset{\overset{}{|}}{CH}}-O-CH_3$
   $O-CH_3$

b. $CH_3CH=N-OH$

c. $CH_3CH=N-NH_2$

d. CH_3CH_2-OH

Question 83

Predict the product obtained when acetic acid, $CH_3-\underset{\underset{O}{\|}}{C}-OH$, is reacted with:

a. isopropyl alcohol

b. isopropylamine?

Answer 83

a. $CH_3-\underset{\underset{O}{\|}}{C}-O-\underset{\underset{CH_3}{|}}{CH}-CH_3$

b. $CH_3-\underset{\underset{O}{\|}}{C}-NH-\underset{\underset{CH_3}{|}}{CH}-CH_3$

Question 84

Predict the major organic product when 2-pentanol is treated with:

a. H_2SO_4/heat

b. $CH_3CH_2-\underset{\underset{O}{\|}}{C}-OH/H^+$

c. $KMnO_4$ (oxidation)

Answer 84

a. $CH_3CH=CHCH_2CH_3$

b. $CH_3CH_2-\underset{\underset{O}{\|}}{C}-O-\underset{\underset{CH_3}{|}}{CH}CH_2CH_2CH_3$

c. $CH_3-\underset{\underset{O}{\|}}{C}-CH_2CH_2CH_3$

Question 85

Cyclohexene is treated with H_2O in the presence of H_2SO_4. The product of that reaction is oxidized. What is the final product obtained?

Answer 85

Cyclohexanone (cyclohexane ring with =O)

Route: Cyclohexene $\xrightarrow{H_2O / H^+}$ Cyclohexanol (–OH) $\xrightarrow{[O]}$ Cyclohexanone (=O)

INDEX

References inside parentheses are for relevant statements (S17, S4, etc.) and questions (Q59, Q82, etc.) containing the term indexed.
Numbers in *italics* refer to illustrations; numbers followed by a (t) indicate tables.

Acetal, formation of, 446 (S17, Q59), 457 (Q82)
Acidity constant (K_a), 330–334 (S5, Q16–Q21), 331(t), 337–38 (S7, Q27, Q28), 341–42 (S9, Q32, Q34), 343 (Q35), 344 (Q37–Q39), 350–51 (Q52, Q53), 352–53 (Q56–Q59), 354 (Q61), 355 (Q62, Q64), 356 (Q66), 358 (Q71)
Acids, combining capacity of, 237 (S7, Q23), 238 (Q25)
　in volumetric analysis, 345 (S11, Q40), 346–47 (Q41–Q43), 359 (Q72, Q73)
　nomenclature of, 97–98 (S16, Q73–Q76), 100 (Q80), 103 (Q90)
　normality of, 238 (Q28), 239 (Q29)
　organic, 429(t), 430 (Q33), 437 (Q44), 438 (S12), 439 (Q48)
　strong, 327 (S3), 328 (Q9–Q12), 349 (Q48), 350 (Q49, Q52), 351 (Q53), 355 (Q63)
　weak, mono-, 330–34 (S5, Q16–Q21), 350 (Q52), 352 (Q56), 353 (Q59)
　poly-, 337–338 (S7, Q27–Q29), 353 (Q58), 356 (Q66)
Activation energy, 265–66 (S4, Q11, Q12), 275 (Q24, Q25)
Activity series, 368 (S5), 369 (Q18), 370 (Q20, Q22), 385 (Q50), 386 (Q52, Q53)
Additions, electrophilic, of alkenes, 441 (S14, Q51), 442 (Q52), 454 (Q76), 457 (Q81)
　of alkynes, 441 (S14), 442 (Q53), 454 (Q76), 457 (Q81)
　nucleophilic, 442 (S15), 443 (Q54), 455 (Q77), 457 (Q82)
Alcohols, acetal formation from, 446 (S17, Q59)
　elimination reactions of, 444–45 (S16, Q57), 456 (Q79), 458 (Q84)
　ester formation from, 446–47 (S17, Q59–Q61), 458 (Q83, Q84)
　ether formation from, 446 (S17, Q59), 447 (Q61), 458 (Q83)
　nomenclature of, 432 (S9, Q37), 433 (Q38), 436 (S11), 437 (Q44, Q46), 452 (Q70), 453 (Q73)
　oxidation of, 447 (S18), 449 (Q63), 455 (Q78), 458–59 (Q84, Q85)
　preparation of, 439 (S13), 440 (Q50), 441 (S14), 447 (S18), 448 (Q62)
　structure of, 429(t), 431 (Q35)

Alcohols *(Continued)*
　substitution reactions of, 439 (S13), 440 (Q49, Q50)
Aldehydes, acetal formation from, 446 (S17, Q59)
　addition reactions of, 442–44 (S15, Q54–Q56), 457 (Q82)
　nomenclature of, 436 (S11), 437 (Q44), 453 (Q72)
　oxidation of, 447 (S18), 449 (Q63), 455 (Q78)
　preparation of, 447 (S18), 449 (Q63), 455 (Q78)
　reduction of, 447–48 (S18, Q62)
　structure of, 429(t), 430 (Q34), 451 (Q69)
Alkanes, formulas for, 410 (S1), 449–50 (Q64–Q66)
　isomerism in, 422 (S5), 424 (Q23), 451 (Q69)
　nomenclature of, 412 (Q4), 413 (Q5), 414(t), 423(t), 450 (Q66), 451 (Q67)
　preparation of, 447–48 (S18, Q62)
　reactions of, 438 (S12, Q47)
　structure of, 410 (S1), 449–50 (Q64–Q66)
Alkenes, addition reactions of, 441–42 (S14, Q51, Q52), 454 (Q76), 459 (Q85)
　formulas for, 413 (S2), 414 (Q9), 450 (Q65)
　geometric *(cis-trans)* isomerism in, 425–26 (S6, Q27–Q29)
　nomenclature of, 413 (S2), 415 (Q10, Q11), 436 (S11), 437 (Q45), 451 (Q67, Q68)
　preparation of, 444–45 (S16, Q57), 456 (Q79), 458 (Q84)
　reduction of, 447–48 (S18, Q62)
　structure of, 413 (S2)
Alkyl groups, 422 (S5), 423(t), 424 (Q24)
Alkyl halides, elimination reactions of, 444–45 (S16, Q57, Q58), 456 (Q79)
　nomenclature of, 432–33 (S9, Q37, Q38), 434–36 (Q39–Q3), 452 (Q70), 453 (Q72)
　preparation of, 438 (S12, Q47), 439–40 (S13, Q49), 441 (S14)
　structure of, 429(t), 430 (Q33)
　substitution reactions of, 440–41 (S13, Q49, Q50)
Alkynes, addition reactions of, 444 (S16), 445 (Q58), 456 (Q79)
　formulas for, 416 (S3), 417 (Q13)

461

462 INDEX

Alkynes *(Continued)*
  nomenclature of, 416 (S3), 417 (Q14, Q15), 436 (S11), 437 (Q45), 453 (Q73)
  preparation of, 444 (S16), 445 (Q58), 456 (Q79)
  reduction of, 447–48 (S18, Q62)
  structure of, 410 (S1), 416 (S3)
Amides, formation of, 446 (S17), 447 (Q60), 458 (Q83)
  structure of, 429(t), 430 (Q34), 454 (Q75)
Amines, amide preparation from, 446 (S17), 447 (Q60)
  nomenclature of, 432–33 (S9, Q37, Q38), 435 (Q40), 453 (Q72)
  preparation of, 439 (S13), 440 (Q49)
  structure of, 429(t), 430–31 (Q34, Q35), 454 (Q75)
Ampholytes, 341 (S9), 342–43 (Q34, Q35), 352 (Q57), 355 (Q62)
Anion, 71 (S4)
Anode, *360,* 362 (S2), 374–77 (S8, Q29A–33), *375,* 378–80 (Q34–Q38), 392 (Q70, Q71)
Antilogarithms, of negative logarithms, 24 (S22, Q69, Q70), 29 (Q91)
  of positive logarithms, 23–24 (S21, Q66–Q68), 29 (Q90, Q92)
Arrhenius equation, 265 (S4), 266 (Q11, Q12)
Atom, definition of, 41 (S1)
  excited state of, 58 (S11, Q41, Q42)
  ground state in, 58 (S11)
  wave-mechanical model of, 46 (S5)
Atomic number, 41 (S1, Q2), 42 (Q3), 59 (Q43–Q45)
Atomic weight, 43–45 (S3, Q6–Q9), 60 (Q47), 115 (Q11, Q12)
Atomic weight unit, 45 (S4, Q9), 46 (Q10)
Aufbau process, 52 (S7), 81 (S10)
Autoionization, 324 (S1)
Avogadro's number, 117 (S6), 118 (Q17, Q18), 126 (Q36, Q37)

Bases, combining capacity of, 237 (S7, Q23), 238 (Q25)
  in volumetric analysis, 345–47 (S11, Q40–Q43), 359 (Q72, Q73)
  normality of, 239–40 (Q28, Q29)
  strong, 329–30 (S4, Q13–Q15), 349 (Q48), 350 (Q50, Q51), 352 (Q55), 355 (Q63)
  weak, mono-, 334 (S6), 335–36 (Q22–Q26), 352 (Q55–Q57), 355–56 (Q64, Q65), 357 (Q68)
  poly-, 339–40 (S8, Q30, Q31), 352 (Q57), 357 (Q67)
Basicity constant (K_b), 335–36 (Q22–Q26), 335(t), 339–41 (S8, Q30, Q31, S9), 342–43 (Q33–Q35), 352 (Q55–Q57), 354 (Q60), 355 (Q62), 356 (Q65), 357 (Q67)
Boiling point, 212–14 (S2, Q4–Q7), 215–16 (S4, Q10, Q11), 217 (Q14–Q17)
  elevation of, 246 (Q44), 248 (S12, Q49), 258 (Q72, Q74), 259 (Q76)
Bond energy, 166 (S7), 168–69 (S8, Q20–Q22), 174–75 (Q35, Q36)

Bonds, pi, 81 (S10), 82 (Q41, Q43), 83–84 (Q44–Q47), 107 (Q102), 413 (S2), 416 (Q12, S3), 418 (Q16), 419 (S4), 420 (Q19), 425 (S6), 429 (S8), 431 (Q36)
  sigma, 81 (S10, Q40), 82–84 (Q43–Q47), 107 (Q102, Q103), 413 (S2)
Boyle's Law, 188–89 (S2, Q3–Q5), 203 (Q32), 204 (Q35)
Buffer solution, 343–44 (S10, Q37–Q39), 358 (Q69–Q71)

Calorimeter, 144 (Q1), 145–47 (S2, Q4, Q5, S3, Q6), 153–54 (Q19–Q21), 164–65 (Q13, Q14), 166 (Q17)
Carboxylic acids, preparation of, 447–48 (S18, Q63), 455 (Q78)
  structure of, 429(t), 430 (Q33), 437 (Q44), 458 (Q83)
Catalysts, 438 (S12)
Cathode, *360,* 362 (S2), 375–77 (S8, Q29A–Q32), *375,* 378–80 (Q34–Q38), 391–92 (Q70, Q71)
Cation, 71 (S4)
Cell diagram, 362–63 (S2, Q4–Q7), 364–65 (Q9, Q10), 388 (Q58, Q59), 390 (Q64)
Cell potential, 363–65 (S3, Q8–Q10), 371–74 (S6, Q23–Q25, S7, Q26–Q28), 388–91 (Q60–Q66)
Cell reaction, 362–63 (S2, Q4–Q7), 378–80 (Q34–Q38), 388–89 (Q58–Q61), 391–92 (Q67–Q71)
Charles' Law, 189–90 (S3, Q6–Q8), 203 (Q33), 204 (Q36)
Chelate, 395 (S2, Q3), 396 (Q5), 406 (Q24)
Clausius-Clapeyron equation, 214–15 (S3, Q8, Q9), 218 (Q18)
Combining capacity, 113–14 (S3, Q7, Q8), 115 (Q10–Q12), 127 (Q40, Q41), 237–38 (S7, Q23–Q26), 383 (Q44)
Common ion effect, 309–11 (S4, Q13–Q16), 319–20 (Q32–Q36), 323 (Q41)
Complex ions, equilibria of, 400 (S5, Q13), 403 (S7)
  formation of, 394 (S1)
Concentration. See *Molarity, Molality,* etc.
  formal, 327 (S3)
Concentration product (Q_c), in precipitation reactions, 311–12 (S5, Q17–Q19), 321 (Q37, Q38), 322 (Q40)
Concentration quotient (Q_c), in gaseous equilibria, 280–81 (S2, Q7–Q9), 295 (Q40, Q41)
Configuration, 422 (S5), 425 (Q27)
Conjugate acid-base pairs, 343–44 (S10, Q37–Q39), 358 (Q69–Q71)
Conservation of Energy, Law of, 146 (S3)
Conservation of Matter, Law of, 132 (S1)
'Conversion factors, 33–35 (S2, Q5–Q7), 38 (Q16, Q17)
Coordination compounds, nomenclature of, 397 (S4), 398–99 (Q9–Q12), 405 (Q21), 406 (Q22, Q25)

INDEX   463

Coordination number, 394 (S1, Q1, Q2), 396 (Q4), 406 (Q22)
Coordination sphere, 396–97 (S3, Q6–Q8), 406 (Q23)
Covalent compounds, formation of, 72–73 (S6, Q20–Q23), 74–76 (Q24, Q26–Q30, S7), 77–84 (S8, Q34–Q36, S9, Q37–Q39, S10, Q40–Q47), 105–8 (Q96–Q106)
  Lewis pictures of, 72 (S6), 73 (Q20, Q22), 74 (Q24, Q25), 75–76 (Q27–Q30, S7, Q31), 103 (Q91–Q93), 105 (Q96, Q97)
  melting points of, 219–20 (S1, Q1–Q3), 221 (Q6, Q7), 225 (Q16, Q17)
  nomenclature of, 96–97 (S15, Q71, Q72), 99 (Q78)
Critical temperature, 215–16 (S4, Q10, Q11), 217 (Q16)
Cycloalkanes, formulas for, 410 (S1), 411 (Q2), 449 (Q64)
  nomenclature of, 412–13 (Q4, Q5), 452 (Q70)
  structure of, 410 (S1)

Dalton's Law, 197–98 (S7, Q22, Q23), 208 (Q48, Q49)
Defining equations, 31–32 (S1, Q1)
Density, 32–33 (Q1–Q4), 35 (Q8), 38 (Q14, Q16), 39–40 (Q20–Q22)
  of gases, 194 (Q16), 198 (S8, Q24), 206 (Q44)
Derivation of equations, 31 (S1), 32–33 (Q2–Q4), 34–36 (Q5–Q9), 37–40 (Q10–Q22)
Diamagnetism, 55 (S9, Q34), 65 (Q66)
Dilution, 240–43 (S8, Q30–Q38), 254–56 (Q62–Q68)
Dipole-dipole bonds, 211 (S1, Q1), 216 (Q12, Q13)
Double bond, 76 (S7), 77 (Q33), 83 (Q44), 107 (Q102), 413 (S2)
Dulong and Petit, Law of, 114–15 (S4, Q9–Q12), 127 (Q41)

$\epsilon°$. See *Reduction potential, standard*
ϵ cell. See *Cell potential*
ΔE. See *Internal energy, changes in*
Effusion, 198–99 (S8, Q24–Q26)
Electrolysis, 374–80 (S8, Q29A–Q32, S9, Q33–Q39), 381 (Q40), 391–92 (Q67–Q72)
Electrolytic cell, 374, *375*
Electron, 41 (S1)
Electron affinity, 66 (S1, Q1, Q2), 100 (Q81, Q82)
Electron configuration, 51–53 (S7, Q24–Q28), 63 (Q57–Q59), 65 (Q65), 68 (S3, Q7), 69–70 (Q9–Q13), 101 (Q85, Q86)
Electron-dot pictures. See *Lewis pictures*
Electron-pair bond, 81–82 (S10, Q40, Q41), 83–84 (Q45–Q47)
Electron-pair repulsion, and molecular geometry, 87–90 (S12, Q54–Q60)
Electron-pair repulsion theory, 87 (S12)
Electronegativity, 91 (S13), 211 (S1)

Electrostatic bond, 70 (S4)
Elements, 41 (S1)
Elimination reactions, 444–45 (S16, Q57, Q58), 456 (Q79)
Empirical formula, 117 (S6), 119–20 (S7, Q21, Q22), 121–23 (S9, Q25–Q28), 128 (Q42–Q44), 129–30 (Q47–Q49), 136 (Q10), 143 (Q27)
Endothermic processes, 144 (S1), 145 (Q3)
Enthalpy, changes in (ΔH), 163–64 (S5, Q10–Q12), 165 (Q14, Q15), 166–69 (S7, Q18, Q19, S8, Q20–Q22), 172–75 (Q29–Q36)
  definition of, 163 (S5)
Enthalpy of formation, 151–53 (S7, Q15–Q18), 157–58 (Q29–Q31), 180–81 (Q10–Q12)
Enthalpy of reaction, 151–53 (S7, Q15–Q18), 157–58 (Q29–Q31), 164–65 (Q13, Q14), 51, 166–69 (S7, Q18, Q19, S8, Q20–Q22), 173–75 (Q32–Q36)
Entropy, 176–78 (S1, Q1–Q3, S2, Q4, Q5), 183–84 (Q15–Q17)
  and spontaneity, 180 (S5)
  changes in (ΔS), 176–78 (Q1–Q5), 180–84 (S5, Q10–Q17), 185–86 (Q21–Q23)
Entropy of reaction, 177–78 (S2, Q4, Q5), 180–84 (Q10–Q17), 185–86 (Q21–Q23)
Equations, chemical, balancing, 132–33 (S1, Q1–Q3), 140 (Q17–Q19)
Equilibrium, acids at, 331 (S5), 337 (S7)
  bases at, 334 (S6), 339 (S8)
  complex ions at, 400 (S5)
Equilibrium, gaseous, 277 (S1), 289 (S7)
  heterogeneous, 289–90 (S7, Q27–Q29)
  liquid-solid, 221 (S3)
  liquid-vapor, 211 (S1), 221 (S3)
  oxidation-reduction and, 360 (S1), 362 (S2)
  precipitates at, 304 (S1)
  solid-vapor, 221 (S3)
Equilibrium concentrations, 282–84 (S4, Q13–Q15), 293 (Q36), 294 (Q37, Q39), 296–97 (Q42, Q43), 298 (Q46)
Equilibrium constants, K_e, gaseous equilibria and, 268 (S5), 277–79 (S1, Q1–Q6), 280–82 (Q7–Q9, S3, Q10–Q12), 283–84 (Q13–Q15), 286 (Q20), 288–91 (Q25, Q26, S7, Q27–Q29, S8, Q30–Q32), 292–95 (Q33–Q41), 296 (Q42), 297–98 (Q44–Q46), 300 (Q48), 301 (Q50), 303 (Q54)
  oxidation-reduction and, 372–74 (S7, Q26–Q28)
  K_p, gaseous equilibria and, 290–92 (S8, Q30–Q32, S9, Q33, Q34), 300–2 (Q49–Q53)
Equilibrium shifts, from changing concentration, 284–86 (S5, Q16–Q21), 297–98 (Q44, Q45)
  from changing volume, 288 (Q25, Q26), 298–99 (Q46, Q47)
Esters, 429(t), 430 (Q33), 431 (Q35)
Ethers, preparation of, 439 (S13), 440 (Q50), 446 (S17, Q59), 447 (Q61)
  structure of, 429(t), 430 (Q33), 432 (Q36), 451 (Q69), 454 (Q75)
Exothermic processes, 144 (S1, Q1)
Exponential notation, adding and subtracting, 16 (S14, Q43, Q44), 26 (Q76, Q77)

Exponential notation *(Continued)*
  dividing, 11 (S9), 12 (Q30), 13–14 (S11, Q36–Q38), 27 (Q80, Q81)
  multiplying, 11 (S9, Q29), 12–13 (S10, Q31–Q35), 27 (Q78), 28 (Q84)
  of large numbers, 4–5 (S3, Q9–Q12), 25 (Q71–Q73)
  of small numbers, 6–7 (S4, Q13–Q16), 25 (Q74)
  raising to a power, 14–15 (S12, Q39, Q40). 27 (Q82)
  significant figures in, 7–8 (S5, Q17–Q19)
  taking roots of, 15 (S13, Q41, Q42)

Faraday's Law, 381–83 (S10, Q40–Q44), 392–93 (Q72–Q74)
First order reactions, 262 (Q3), 263 (Q7), 264–65 (S3, Q9), 266 (Q12), 270 (Q16), 273 (Q20)
Formula, of ionic compounds, 70–72 (S4, Q14, Q15, S5, Q16–Q19), 102–3 (Q87–Q90)
Formula units, 117 (S6), 118 (Q17, Q18), 126 (Q36, Q37)
Formula weight, 119–20 (S7, Q21, Q22)
Free energy, 178 (S3)
  changes in, 178–82 (S3, Q6–Q7, S4, Q8, Q9, S5, Q10–Q14), 184–86 (Q18–Q23)
  and K_p, 292 (S9, Q33, Q34), 302–3 (Q52–Q54)
Free energy of formation, 178–89 (S3, Q6, Q7, Q8, Q9), 184 (Q18), 185 (Q20), 292 (S9, Q33, Q34), 302 (Q53)
Free energy of reaction, 178–82 (S3, Q6, Q7, Q8, Q9, S5, Q10–Q13), 184 (Q18), 185–86 (Q20–Q23)
Freezing point, lowering of, 245–48 (S11, Q43–Q48, S12), 249 (Q50), 258–59 (Q72–Q75), 260 (Q77)
Functional groups, 429 (S8), 429(t), 434(t), 436 (S11), 454 (Q75)

ΔG. See *Free energy, changes in*
$\Delta G°_f$. See *Free energy of formation*
ΔG_{rx}. See *Free energy of reaction*
Galvanic cell, 360 (S1), *360,* 361 (Q2), 364 (Q8, Q9), 371 (S6, Q24)
Gas, ideal, 200 (S9)
  real, 200 (S9)
Gaseous state, 187 (S1)
Geometry, minimum repulsion, 87–89 (S12, Q54–Q58), 109 (Q109, Q110)
  molecular, 84–91 (S11, Q48–Q53, S12, Q54–Q60, S13), 108 (Q105), 109 (Q107–Q110), 420 (Q17)
Gibbs-Helmholtz equation, 180–82 (S5, Q10–Q14), 185–86 (Q21–Q23)
Graham's Law, 198–99 (S8, Q24–Q26), 208–9 (Q50, Q51)
Gram atomic weight (GAW), 111 (S1, Q1, Q2), 113–15 (S3, Q7, Q8, S4, Q9–Q12), 120 (S8), 122 (Q25), 124 (Q29), 126 (Q35), 127 (Q40), 382 (Q41, Q42), 392–93 (Q72, Q73)

Gram equivalent weight (GEW), 112–14 (S2, Q4–Q6, S3, Q7, Q8), 115 (Q10–Q12), 126–27 (Q38, Q39), 237 (S7), 238 (Q25, Q26), 382–83 (Q43, Q44)
Gram molecular weight (GMW), 117–18 (S6, Q15–Q17), 124 (Q30)
  and root mean square velocity, 202 (S10, Q29, Q30)
  determination by vapor density method, 194 (S5, Q15), 195 (Q17), 206 (Q43)
  from lowering freezing point, 246 (Q45), 259 (Q75)
  in Graham's Law, 198 (S8), 199 (Q25, Q26), 209 (Q51)

ΔH. See *Enthalpy, changes in*
ΔH_{atom}. See *Heat, of atomization*
ΔH_{comb}. See *Heat, of combustion*
ΔH_{diss}. See *Heat, of dissociation*
$\Delta H°_f$. See *Enthalpy of formation*
ΔH_{fus}. See *Heat, of fusion*
ΔH_{rx}. See *Enthalpy of reaction*
ΔH_{vap}. See *Heat, of vaporization*
Half-life, 264–65 (S3, Q9–Q10), 274 (Q23)
Halogenation, of alkanes, 438 (S12, Q47), 456 (Q80)
  of aromatic hydrocarbons, 438 (S12, Q47), 456 (Q80)
Heat
  of atomization, 166 (S7), 167 (Q19), 174 (Q34)
  of combustion, 145–46 (Q4, Q5), 153–54 (Q19–Q21), 165 (Q15, Q16)
  of dissociation, 166–67 (S7, Q18, Q19), 173–74 (Q33, Q34)
  of formation, 151–53 (S7, Q15–Q18), 157–58 (Q29–Q31), 180–81 (Q10–Q12)
  of fusion, 219 (S1)
  of reaction, 151–53 (S7, Q15–Q18), 157–58 (Q29–Q31), 164–65 (Q13, Q14), 168–69 (S8, Q20–Q22), 173–75 (Q32–Q36), 180–82 (Q10–Q13), 185–86 (Q21–Q23)
  of vaporization, 211 (S1), 212 (Q2, Q3), 214–15 (S3, Q8, Q9), 216 (Q13), 217 (Q15, Q17), 218 (Q18)
Heat capacity, 145–46 (S2, Q4, Q5), 153–54 (Q19–Q21)
Heat flow, 144 (S1, Q2), 145 (S2, Q3, Q4), 147–48 (Q6, S4, Q7–Q9), 149 (Q10, Q11), 150–51 (Q12–Q14), 153–57 (Q19–Q28), 159 (S1, Q1, Q2), 161 (Q6), 162–63 (Q8, Q9), 164–66 (S6, Q13–Q17), 170 (Q23), 171 (Q27, Q28), 172–73 (Q31, Q32)
Heisenberg uncertainty principle, 46 (S5)
Henry's Law, 243–44 (S9, Q39, Q40), 257 (Q69, Q70)
Hess's Law, 150–51 (S6, Q12–Q14), 156–57 (Q27, Q28), 166–67 (S7, Q18, Q19)
Hückel Rule, 419 (S4), 421–22 (Q20, Q21), 454 (Q74)
Hund's Rule, 53 (S8)
Hybridization of orbitals, 79–80 (S9, Q37–Q39), 105–6 (Q98, Q99)

INDEX 465

Hybridization of orbitals *(Continued)*
and molecular geometry, 84–87 (S11, Q48–Q53)
in alkanes, 412 (Q3)
Hydrocarbons, acyclic, 410 (S1)
aliphatic, 410–13 (S1, Q1–Q5)
aromatic, formulas for, 418 (S4)
isomerism in, 422 (S5)
nomenclature for, 418 (S4), 427–28 (S7, Q30–Q32), 452 (Q71)
reactions of, 438–39 (S12, Q47, Q48), 456 (Q80)
structure of, 418 (S4), 420 (Q18), 421–22 (Q20, Q21), 454 (Q74)
cyclic. See *Cycloalkanes*
Hydrogen bonds, 211 (S1, Q1), 212 (Q3), 216 (Q12, Q13)

Ideal gas, definition of, 200 (S9)
Ideal gas equation, 191–93 (S4, Q9–Q14), 204 (Q34), 205–6 (Q37–Q43)
Ideal solutions, 244 (S10)
Instability constants, 400–1 (S5, Q13–Q15), 402–4 (Q16–Q18, S7, Q19, Q20), 407–9 (Q26–Q31)
Instability constants, values for, 400(t)
Intermediates, in step-wise reactions, 267 (S5)
Intermolecular forces, 211 (S1), 216 (Q12)
Internal energy, changes in (ΔE), 159 (S1, Q1, Q2), 161 (Q6), 162–65 (Q8, Q9, S5, Q10–Q12, S6, Q13–Q16), 170 (Q23), 171–73 (Q27–Q32)
Ion product constant of water (K_w), 324–25 (Q1–Q3), 326 (Q5, Q7), 328 (Q11–Q12), 329–30 (Q13–Q15, S5), 332–33 (Q17–Q19), 334 (S6), 335–36 (Q23, Q24), 338 (Q28–Q29), 340–41 (Q31, S9), 342–43 (Q33–Q36, S10), 348 (Q44–Q48), 351–52 (Q53–Q57), 354–57 (Q60–Q67)
Ion-electron method, 365 (S4)
Ionic compounds, 66 (S1), 70–72 (S4, Q14, Q15, S5, Q16–Q19), 102 (Q87–Q89)
melting points of, 220 (S2, Q4, Q5), 221 (Q6, Q7), 224 (Q14), 225 (Q16)
nomenclature of, 94 (S14, Q66), 94(t), 95–96 (Q67–Q70), 99 (Q77–Q79), 102–3 (Q88–Q90)
solubility of, 305(t)
Ions, electron configuration of, 56–58 (Q35–Q40), 64 (Q63, Q64)
formation of, 56 (S10), 66 (S1), 67–68 (S2, Q3–Q5, S3, Q7), 69–70 (Q8–Q13), 101 (Q83–Q86)
polyatomic, 75 (Q27, Q28), 76 (Q31), 94(t)
Isomers, *cis-trans*, 425 (S6), 426 (Q28, Q29), 451 (Q68)
stereo-, 422 (S5), 425 (Q27)
structural, 422 (S5), 425 (Q27)
of alkanes, 423–24 (Q22, Q23), 451 (Q69)
of alkenes, 424–25 (Q25, Q26)
of substituted benzene compounds, 427–28 (S7, Q30–Q32)
Isotopes, 42 (S2, Q4), 43–45 (Q5–Q8, S4), 46 (Q10), 59–60 (Q46–Q49)

K_a. See *Acidity constants*
K_b. See *Basicity constants*
K_c. See *Equilibrium constants*
K_{inst}. See *Instability constants*
K_p. See *Equilibrium constants*
K_{sp}. See *Solubility product constant*
Ketones, 429(t), 430 (Q34), 451 (Q69)
addition reactions of, 442–43 (S15, Q54–Q55)
preparation of, 447 (S18), 455 (Q78)
reduction of, 447 (S18)
Kinetics, 261 (S1)

Law of Chemical Equilibrium, 261 (S1), 277 (S1)
Law of Mass Action, 261 (S1)
Lewis pictures, of ions, 67–68 (S2, Q3–Q6), 101 (Q83, Q84), 105 (Q95)
of molecules, 72–73 (S6, Q20, Q22), 74–76 (Q24–Q30, S7, Q31), 103 (Q91–Q93), 105 (Q96, Q97)
Ligands, 394 (S1, Q1, Q2), 396 (Q4), 406 (Q23)
names of, 398(t)
Limiting reagent, 138–39 (S5, Q13, Q14, Q16), 143 (Q26)
Liquid state, 211 (S1)
Logarithms, common, 17–19 (S15, Q46–Q53), 28 (Q83, Q85)
natural, 22–23 (S20, Q64, Q65), 28 (Q86)
of a number raised to a power, 21 (S18, Q60, Q61), 29 (Q88)
of products, 19–20 (S16, Q54–Q56), 28 (Q87)
of quotients, 20–21 (S17, Q57–Q59), 28 (Q87)
of the root of a number, 22 (S19, Q62, Q63), 29 (Q89)
Lone-pair electrons, 87–89 (S12, Q54–Q56), 90–91 (Q59, Q60, S13), 104 (Q94), 105 (Q96, Q97)

Mass number, 41–43 (S1, Q1–Q5), 59 (Q43–Q45)
Melting point, 219–221 (S1, Q1–Q3, S2, Q4–Q7, S3), 223 (Q10), 224–25 (Q12–Q16), 226 (Q19)
Molality, 235–36 (S6, Q19–Q22), 254 (Q61)
Molarity, 232–34 (S5, Q12–Q18), 252–54 (Q56–Q60)
Mole, 117 (S6, Q16), 118–19 (Q17–Q20), 125 (Q32–Q34), 126 (Q36, Q37)
Mole fraction, 230–32 (S4, Q8–Q11), 251 (Q55), 253 (Q59)
Molecular formula, 117 (S6), 118 (Q17), 119–20 (S7, Q21, Q22), 123 (Q28), 128 (Q42–Q44), 130 (Q49)
Molecular orbital theory, 80–84 (S10, Q40–Q47)
Molecular weight, 116 (S5, Q13, Q14), 119 (S7), 124 (Q31)
Molecules, 72–78 (S6, Q20–Q30, S7, Q32, Q33, S8, Q34–Q36), 103 (Q91–Q93)
geometry of, 84–91 (S11, Q48–Q53, S12, Q54–Q60, S13), 108–9 (Q106–Q110)

Molecules *(Continued)*
  polarity of, 91 (S13), 92–93 (Q63–Q65), 110 (Q111), 211 (S1)

Nernst equation, 371–72 (S6, Q23–Q25), 389–90 (Q62–Q66)
Net ionic equations, 305 (S2), 306 (Q3–Q7), 317 (Q26)
Neutron, 41–42 (S1, Q1–Q3), 59 (Q43–Q45)
Nitration, of aromatic hydrocarbons, 438 (S12), 439 (Q48), 456 (Q80)
Nomenclature, IUPAC, 433–37 (S10, Q39–Q43, S11, Q44–Q46), 438 (Q47), 451 (Q67, Q68), 453 (Q73)
  of acids, 97–100 (S16, Q73–Q80)
  of alkanes, 412–13 (Q4, Q5), 414(t)
  of alkenes, 413 (S2), 415 (Q10, Q11)
  of coordination compounds, 397–99 (S4, Q9–Q12), 405 (Q21), 406 (Q22, Q25)
  of covalent compounds, 96–97 (S15, Q71, Q72), 99 (Q78)
  of ionic compounds, 94 (S14, Q66), 95–96 (Q67–Q70)
Normality, 237 (S7), 239–40 (Q27–Q29), 252–53 (Q56–Q58), 254 (Q60)
Nucleus, 41 (S1)

OEIOU (one equation in one unknown), 31 (S1), 32–33 (Q2–Q4), 37–40 (Q13–Q22)
Orbitals, antibonding, 81 (S10)
  See also *Orbital diagrams, molecular, M-O*
  atomic, 46 (S5), 49 (Q21)
  See also *Orbital diagrams, of atoms and ions*
  bonding, 81 (S10)
  See also *Orbital diagrams, molecular, M-O*
  molecular, 81 (S10)
  See also *Orbital diagrams, molecular*
Orbital diagrams, molecular, M-O, 81–83 (S10, Q40, Q41, Q43, Q44), 107–8 (Q102–Q104)
  V-B, 77–78 (S8, Q34–Q36), 79–80 (Q37–Q39), 85 (Q48), 105–6 (Q98–Q101)
  of atoms and ions, 53–55 (S8, Q29–Q32), 63–64 (Q60–Q64)
Orbital overlap, 77 (S8)
Order of reaction, 263–64 (S2, Q3–Q6), 272 (Q19)
  determination of, 270–271 (S6, Q16, Q17), 273 (Q20, Q21)
Osmosis, definition of, 249 (S13)
Osmotic pressure, 249 (S13), 250 (Q52, Q53), 258 (Q73), 260 (Q78)
Overvoltage, 377 (S9)
Oxidation, definition of, 362 (S2)
  of alcohols, 447 (S18), 449 (Q63), 455 (Q78), 458–59 (Q84, Q85)
  of aldehydes, 447 (S18), 449 (Q63), 455 (Q78)

Oxidation-reduction equations, 365–67 (S4, Q11–Q14), 386–87 (Q54–Q57)
Oxidizing agents, 368 (S5, Q15, Q16), 369 (Q19)

Paramagnetism, in atoms and ions, 54 (S9), 55 (Q33, Q34), 65 (Q66)
  in molecules, 83 (Q44), 107 (Q102, Q103)
Pauli exclusion principle, 53 (S8)
Per cent by volume, 230 (S3, Q6, Q7), 251 (Q55)
Percentage composition, 120–22 (S8, Q23, Q24, S9, Q25), 129 (Q45–Q47), 130 (Q49), 136 (Q10), 143 (Q27)
Percentage yield, 137 (S4, Q11, Q12), 142 (Q24, Q25)
pH, 325–26 (S2, Q4–Q7), 334 (Q21), 337 (Q27), 340 (Q30), 342–43 (Q32–Q36), 344 (Q39), 348–49 (Q44–Q47), 354–55 (Q60–Q62), 358 (Q70, Q71)
Phase diagrams, 221–24 (S3, Q8–Q12), 225–27 (Q17–Q22)
Polar bond, 91–92 (S13, Q61, Q62), 429 (S8)
Precipitation, equilibrium in, 304 (S1, Q1, Q2), 307 (S3), 309 (S4), 403 (S7)
  in gravimetric analysis, 314–15 (S6, Q22, Q23), 323 (Q43)
  in volumetric analysis, 315–16 (S7, Q24, Q25), 323 (Q42)
  predictions from Q^c, 311–13 (S5, Q17–Q21), 321–22 (Q37–Q40)
Precision, 2 (S2), 3 (Q6, Q7), 26 (Q75)
Pressure, of gases, 187 (S1, Q1), 188–89 (S2, Q3–Q5), 191–93 (S4, Q9–Q14), 197–98 (S7, Q22, Q23), 203 (Q31, Q32), 204 (Q34–Q36)
Protons, 41–42 (S1, Q1–Q3)

Quantum number, and spectroscopic notation, 51 (S6, Q22, Q23)
  angular momentum, 46–48 (S5, Q11–Q16), 49 (Q21), 61 (Q50, Q52)
  magnetic, 46 (S5), 47–48 (Q14–Q18), 49 (Q21), 62 (Q54, Q55)
  principal, 46–47 (S5, Q11–Q13), 49 (Q19–Q21), 61–62 (Q50–Q53)
  spin, 46 (S5), 48 (Q17, Q18), 49 (Q21), 62 (Q54)

Radius, ionic, 220–21 (S2, Q4–Q7)
Raoult's Law, 244–45 (S10, Q41), 257 (Q72)
Rate constant, 262–63 (S1, Q1, Q2, S2), 263–64 (Q7, Q8), 270–71 (Q16, Q17), 273 (Q20, Q21), 275 (Q24, Q25)
Rate equation, 261 (S1, Q1, Q2), 262–64 (Q3–Q8), 268–70 (Q13–Q15, S6), 272 (Q18)
Rate-determining step, 267–69 (S5, Q13–Q15), 275–76 (Q26, Q27)
Reaction, extent of, 281–82 (S3, Q10–Q12), 294 (Q38)

Reaction *(Continued)*
　predicting direction of, 280 (S2, Q8), 281 (Q9)
Reaction mechanism, 267–69 (S5, Q13–Q15), 275–76 (Q26, Q27)
Reactivity, in organic molecules, 429 (S8), 454–55 (Q76, Q77)
Real gases, definition of, 200 (S9)
Reducing agents, 368 (S5), 369 (Q17, Q18), 370 (Q20–Q22), 385 (Q48, Q50), 386 (Q52, Q53)
Reduction, definition of, 360 (S1), 362 (Q3, S2)
　of aldehydes and ketones, 447–48 (S18, Q62)
　of alkenes, 447–48 (S18, Q62)
　of alkynes, 447–48 (S18, Q62)
Reduction potential, standard, and affinity for electrons, 360–62 (S1, Q1–Q3), 383–84 (Q45–Q47)
　and electrolysis, 377–79 (S9, Q33–Q37)
　and oxidizing strength, 368 (S5, Q15, Q16), 369–70 (Q19, Q20), 385 (Q50)
　and reducing strength, 368 (S5), 369 (Q17, Q18), 370 (Q21, Q22), 385 (Q49), 387 (Q56)
　values, 361(t)
Representative elements, 66 (S1), 68 (S3)
Resonance, in aromatic hydrocarbons, 419 (S4)

ΔS. See *Entropy, changes in*
$\Delta S°_{rx}$. See *Entropy of reaction*
Salts, combining capacity of, 237 (S7, Q23), 238 (Q26)
　normality of, 239 (Q27)
　of strong acids and strong bases, 341 (S9), 343 (Q36), 354 (Q60, Q61)
　of strong acids and weak bases, 342–43 (S9, Q32), 354 (Q61)
　of weak acids and strong bases, 341 (S9), 342 (Q33), 343 (Q35), 352 (Q55, Q57), 354 (Q60), 356 (Q65), 357 (Q67)
　of weak acids and weak bases, 341 (S9), 342 (Q34)
Schrödinger equation, 46 (S5)
Second order reaction, 262 (Q4), 264 (Q8, S3), 265 (Q10), 266 (Q11), 271 (Q17), 273 (Q21)
Significant figures, determination of, 2–3 (S2, Q3–Q5), 8 (Q19)
　in addition and subtraction, 10–11 (S8, Q26–Q28), 26 (Q76, Q77), 27 (Q79)
　in exponential notation, 7–8 (S5, Q17–Q19)
　in multiplication and division, 8–10 (S6, Q20–Q23, S7, Q24, Q25)
　precision and, 2 (S2), 3 (Q4, Q6)
Solid state, 219 (S1)
Solubility, by complex formation, 403–4 (S7, Q19, Q20), 408–9 (Q30, Q31)
　of gases, 243–44 (S9, Q39, Q40), 257 (Q69, Q70)
　of ionic compounds, and K_{sp}, 307–9 (S3, Q8–Q12), 317–19 (Q27–Q31)
　rules for, 305(t)

Solubility product constant (K_{sp}), 307 (S3, Q8), 307(t), 308–9 (Q9–Q12), 310–11 (Q14–Q16, Q17), 312 (Q18, Q19), 318–19 (Q28–Q31), 321–22 (Q37–Q41), 403–4 (S7, Q19, Q20), 408–9 (Q30, Q31)
Solute, 228 (S1, Q2)
Solutions, 228 (S1)
Solvent, 228 (S1, Q1), 229 (Q3), 251 (Q54)
Specific heat, 114 (S4, Q9), 115 (Q12), 127 (Q41)
Spectroscopic notation, 51 (S6, Q22, Q23), 61 (Q51), 62 (Q56)
Spontaneity, 179–80 (S4, Q8, Q9, S5), 182 (Q14), 185 (Q19), 186 (Q20, Q22)
Stoichiometric statement, 133–34 (S2, Q4, Q5), 135–36 (Q6–Q8), 140–41 (Q20, Q21)
　with gases, 195–96 (S6, Q18–Q21), 207 (Q45–Q47)
Sublimation, 221 (S3)
Substitution, electrophilic, 438–39 (S12, Q47, Q48), 456 (Q80)
　nucleophilic, 439–40 (S13, Q49, Q50)
Sulfonation, of aromatic hydrocarbons, 438 (S12), 439 (Q48)
Sulfonic acids, 438 (S12), 439 (Q48)

Temperature, 36–37 (S3, Q10–Q12), 39 (Q18)
　of gases, 187 (S1, Q2), 189–91 (S3, Q6–Q8, S4, Q9), 202–3 (S10, Q29–Q31, Q33, Q34)
Theoretical yield, 135 (S3, Q6, Q7), 136 (Q9), 141 (Q22), 142 (Q24)
Thermodynamics, first law of, 159 (S1)
　second law of, 176 (S1)
Third order reaction, 263 (Q5), 268 (Q13, Q14), 272 (Q19)
Triple bond, 76 (S7, Q31), 77 (Q33), 82 (Q43), 83 (Q45), 84 (Q47), 105 (Q97)
Triple point, 221 (S3), 223 (Q10), 227 (Q21)
Trouton's Rule, 213 (S2, Q4–Q6), 217 (Q15, Q17)

Valence-bond theory, and bonding, 77–79 (S8, Q34–Q36, S9), 105–6 (Q98, Q99)
　and geometry, 84–87 (S11, Q48–Q53)
Values, defined and measured, 1 (S1, Q1, Q2), 4 (Q8), 25 (Q71)
van der Waals equation, 200–1 (S9, Q27, Q28), 209 (Q52)
van der Waals forces, 211–12 (S1, Q1, Q2), 216–17 (Q12–Q14)
van't Hoff factor, 248–49 (S12, Q49, Q50), 259–60 (Q76, Q77)
Vapor density method, 194–95 (S5, Q15–Q17), 206 (Q43)
Vapor pressure, 214–15 (S3, Q8, Q9), 216 (Q13), 217 (Q15, Q17), 218 (Q18), 221 (S3)
　lowering of, 244 (S10), 245 (Q42), 258 (Q73)
Velocity, root mean square, 202 (S10, Q29, S30), 209–10 (Q53, Q54)

Volume, of gases, 188–90 (S2, Q3–Q5, S3, Q6, Q7), 191–93 (S4, Q9–Q14), 200–1 (S9, Q27, Q28), 203–6 (Q32–Q37), Q39–Q42), 209 (Q52)

Weight per cent, 229 (S2, Q4, Q5), 251 (Q55)
Work, in changes of state, 159–61 (S1, Q1, Q2, S2, Q3–Q6), 162 (Q8), 163–65 (Q9, Q10–Q14, Q16), 170–73 (Q23–Q32)

Work *(Continued)*
   pressure-volume, 160 (S2, Q3, Q4), 161 (Q5, Q6), 162–64 (Q8, Q9, S5, Q10–Q12), 165 (Q14), 170–73 (Q24–Q32)
   unavoidable, 178 (S3)
   useful, 178 (S3)

Zeroes, as significant figures, 2–3 (S2, Q3–Q5), 8 (Q19), 25 (Q73, Q74)